# THE OLIVE: ITS PROCESSING AND WASTE MANAGEMENT

# FOOD SCIENCE AND TECHNOLOGY SERIES

**Food Science and Technology: New Research**
*Lorenzo V. Greco and Marco N. Bruno (Editors)*
2008. ISBN: 978-1-60456-715-1

**Food Science and Technology: New Research**
*Lorenzo V. Greco and Marco N. Bruno (Editors)*
2008. ISBN: 978-1-61668-106-7 (Online Book)

**The Price of Food**
*Meredith N. Fisher (Editor)*
2009. ISBN: 978-1-60692-440-2 (Softcover Book)

**Food Processing and Engineering Topics**
*Maria Elena Sosa-Morales and Jorge F. Velez-Ruiz (Editors)*
2009. ISBN: 978-1-60741-788-0

**Traditional Chinese Foods: Production and Research Progress**
*Li Zaigui and Tan Hongzhuo*
2009. ISBN 978-1-60692-902-5

**Food Science Research and Technology**
*Isaak Hülsen and Egon Ohnesorge (Editors)*
2009. ISBN 978-1-60741-848-1

**Food Hydrocolloids: Characteristics, Properties and Structures**
*Clarence S. Hollingworth (Editor)*
2010 ISBN: 978-1-60876-222-4

**The Olive: Its Processing and Waste Management**
*José S. Torrecilla*
2010. ISBN: 978-1-60876-719-9

FOOD SCIENCE AND TECHNOLOGY SERIES

# THE OLIVE: ITS PROCESSING AND WASTE MANAGEMENT

## JOSÉ S. TORRECILLA

Nova Science Publishers, Inc.
*New York*

LIBRARY OF CONGRESS CATALOGING-IN-PUBLICATION DATA
Torrecilla, Josi S.
  The olive : its processing and waste management / Josi S. Torrecilla.
     p. cm.
  Includes index.
  ISBN 978-1-60876-719-9 (hardcover)
  1. Olive oil. 2. Olive oil industry--Refuse and refuse disposal.  I. Title.
  TP683.T67 2009
  664'.362--dc22
                                      2009052745

*Published by Nova Science Publishers, Inc, ✛ New York*

# CONTENTS

**Preface**                                                            vii

**Section 1. Olive Tree**

**Chapter 1**      The History of the Olive Tree                         3

**Chapter 2**      The Life of the Olive Tree                           11

**Section 2. Table Olives**

**Chapter 3**      Table Olives                                         25

**Section 3. Olive Oil**

**Chapter 4**      Industrial Methods to Produce Extra Virgin Olive Oil  47

**Chapter 5**      Virgin Olive Oil Quality                             65

**Chapter 6**      Industrial Methods to Produce Refined Olive Oil      83

**Section 4. Waste from the Olive Oil Sector**

**Chapter 7**      Waste and its Management in the Olive Sector         95

**Index**                                                             123

# PREFACE

The Olive, Its Processing and Waste Management is a comprehensive resource for commercial growers, farmers, students and researchers. Throughout the book, to give depth to the knowledge of each topic, current and relevant scientific references can be found. At the same time, this book combines historical, agricultural and industrial matters related with the olive, i.e., tips and explanations relating to the olive tree and its harvesting and pruning and the methods to produce extra virgin olive oil, refined olive oil and table olives can be found. In the last part of the book, an important section dealing with the wastes generated and their management is added. In this section, classical and new methods (applied or in development) from the research or industrial spheres are studied.

This book "The Olive, Its Processing and Waste Management", endeavors to cover a broad spectrum regarding different aspects of the olive oil sector. It can be classified into four main sections viz. olive tree, table olives, manufacture of extra virgin olive oil and other low-grade olive oils and treatment and management of their wastes. The first covers the description of the main characteristics of the olive tree, types, growth, cultivation, and practical advice on fruit harvesting, harvesting time, industrial productivity, *etc*. The second group deals with a historical description of the main characteristics of table olives such as varieties, health and nutritional characteristics, *etc*. Then, different methods (from squeezing the olives to the present day two-phase method) used to produce extra virgin olive oil will be shown and compared from four points of views *viz*. health features, environmental impact, extra virgin olive oil quality and economic yield. In addition, in this section, the industrial methods to produce low-grade olive oil have been described. In the final section, the environmental impact of the hazardous chemicals generated in the waste, taking into consideration the raw material, processing methods, *etc.* have been detailed. Finally, in order to manage this waste in the best possible way, classical and novel methods have been described.

The author is grateful to Ian Ure (my Scottish teacher of English) for helping me to dig into the marvelous language that is English and for revising this book.

.

*Eben-Ezer*

This book is dedicated to my parents, wife and daughter. My wife, Teresa, has always given me the support to pursue my dreams and our daughter, Irene, who is a marvelous joy in our life.

# SECTION 1. OLIVE TREE

*Chapter 1*

# THE HISTORY OF OLIVE TREE

## ABSTRACT

Fossil Leaves of the *oleaster* fruit from *circa* sixty millenniums ago have been found on the Island of Santorini in Greece which demonstrates the length of time that the olive tree has been in the world. In addition, in the light of fossils found, it can be assumed that fifty millenniums ago, the *oleaster* fruit (olives in their wild state) would occasionally be collected along with other wild edible fruits to supplement the daily diet of the inhabitants. As a consequence of maritime trade routes, knowledge of the olive tree spread throughout the world and the number of its applications also increased notably. In this Chapter, the history of the olive tree is treated briefly.

**Keywords:** History of the olive tree; Olive; Olive tree.

## 1. THE OLIVE HISTORY

The olive tree is one of the world's oldest cultivated trees (Kapellakis *et al.*, 2008). According to Greek mythology, two Greek gods Athena, the goddess of wisdom, and Poseidon, the god of the sea, wanted to be the patron of Attica, the section of Greece that included the city of Athens. The other gods on Mount Olympus devised a contest for them, specifying that the winner would be the one who could provide the best gift to the people of Attica. Poseidon struck the ground with his trident and a horse sprang forth; Athena did likewise with her spear, and an olive tree grew. The gods decided that the olive tree, as a symbol of peace and agriculture, was a much better gift than Poseidon's horse, a symbol of war. And so, Athena became the patron of Attica, and its principal city, the city of the olive tree. Athens was named after her and according to Greek mythology, the goddess of wisdom, Athena, taught people how to use the olive tree, giving rise to the olive tree as an important symbol of peace, wisdom and victory for Greece and later between nations. The ancient Egyptians believed that the goddess Isis transmitted the knowledge of the growing and usage of olives to the people.

Apart from mythology, there are many items related with the olive throughout history. In addition it is one of the plants most cited in the literature. The olive oil was dubbed by Homer as "liquid gold" and the olive also appeared in his book titled the Odyssey written *ca.* 800 BC. Referring to diet, Horace stated "as for me, olives, endives and smooth mallows provide sustenance". On 1779, James Burnett, Lord Monboddo wrote that the olive oil was one of the most perfect foods in the daily diet.

In ancient Greece, athletes ritually rubbed their bodies all over with olive oil, and from about 700 BC, the winners of Olympic Games were awarded a wreath of olive branches (Kapellakis *et al.*, 2008). The ancient Egyptians crowned their dead with olive branches (Boskou, 2006). Olive oil represented emblems of benediction and purification, and the Egyptians also ritually offered it to deities and powerful figures; some were even found in Tutankhamen's tomb. For thousands of years, the Mediterranean peoples have considered olive oil as sacred. Its inhabitants considered that olive oil was more than mere food, having medicinal and magical properties, an endless source of fascination and wonder and the fountain of great wealth and power.

The Bible mentions olives and olive trees in many places (nearly 200), both in the Old and New Testaments, where it was used as fuel for lamps, anointment, a means of payment, *etc.* In the Old Testament, Noah released a dove, which returned holding an olive branch as a sign of the *re*-establishment of peace between God and humanity (Genesis 8, 10-11). The first recorded evidence of olive oil extraction comes from the Hebrew Bible and took place during the Exodus of the Jews from Egypt, during the thirteenth century BC. During this time, the oil was produced by hand-squeezing the berries and stored in special containers under the guard of priests. The term Messiah literally means "anointed (one)". Olive oil has religious symbolism for healing and strength and also for the purpose of consecration.

In relation with the economics of the olive tree, one of the most reliable proofs of the importance of the olive oil trade in those times consists of the numerous stirrup-jars used exclusively to transport olive oil throughout the Mediterranean Sea. Depending on the region and habits of the population, containers such as amphorae, skins, barrels, jars, *etc.* were used as a measure of olive oil and wine (Boskou, 2006). As an example, equipment to measure the quantity of olive oil in Roman times is shown in figure 1

Figure 1. Vessel used in Roman times to quantify olive oil.

Thanks to its properties, throughout history, the olive tree and its oil were used not only as an important part of the inhabitant's daily diet, but also as main source of lighting, to manufacture wooden items or as a substitute for soap in the cleansing of the human body. Due to the hydration and thermal characteristics of the olive oil, since ancient times, the olive oil has been used as an ingredient in beauty and body care products. There were many tips to maintain healthy hair that consisted of treating it with olive oil (Boskou, 2006). Traditionally, olive oil has been used as a medicine and in particular, as an all purpose antiseptic or to cure small wounds, skin irritations, rheumatism, earache and abdominal pain (Boskou, 2006).

In recent centuries, from an environmental point of view, as a result of the manufacturing processes of olive oil (its waste) has been used as food for domestic animals, fertilizer, a source of power, *etc.* (Torrecilla, 2001). In the second half of the twentieth century, the seed-oil has also been industrially used as a component in the manufacture of cosmetics and soap (Boskou, 2006).

## 2. PAST AND PRESENT OF THE OLIVE TREE

The olive tree is a long living evergreen tree. Depending on its age, its trunk size, rough bark texture and shape vary greatly. As an example, young and older trunks are shown in figure 2. In general, the term Olive is used to describe the *Olea Europaea;* this word is also used for more than 35 species of evergreen shrubs and trees of the genus *Olea* in the olive family. There are many varieties of olive which exhibit minor phenotypical and genetic differences, that is, most of the differences between them are based on the size, color, oil content, fatty acid composition, *etc.* (Boskou, 2006). In some cases, the nomenclature of *olea europaea L. sativa* is used to distinguish it from the wild olive subspecies *oleaster.* In general, the olive tree is mainly found in warm temperate regions similar to those of the Mediterranean area.

Figure 2. Olive tree trunks; (a) young plant from Karkur, Israel; (b) Older trunk from Lisbon, Portugal.

A controversial topic is the ancestor of olive tree. It is believed that *olea chrysophylla* found in Asia and Africa is the progenitor of the olive tree. Some scientists consider that the *olea oleaster* is the ancestor of the Mediterranean wild olive, while still others believe that it is an intermediate in the development from *olea chrysophylla* to *olea europaea* (Boskou, 2006). In any case, most varieties of olives trees come from Greece, Italy, Palestine and Syria. And although some of them are more widely distributed than others, the present large number of varieties is due to the fact that the cultivation of olive trees began a long time ago (more than 40,000 years) (Boskou, 2006). This gives rise to another conflictive question which is the subject of much debate and will be illustrated here. Where did olive trees originate and when?

In the light of fossils found in Italy, the origin of the olive tree's ancestor dates back to the Tertiary period (1 million years ago) (Kapellakis *et al.*, 2008). Much later, between 50,000 - 60,000 years ago, the oldest fossilized leaves of *oleaster* fruit were found on the island of Santorini, Greece. In the Neolithic age, *circa* 8,000 BC, the *oleaster* fruit (wild forms of olive) would occasionally be collected with other wild edible fruits to supplement the daily diet (Lumaret *et al.*, 2004; Boskou, 2006). *Circa* 6,000 years ago, it is believed that the cultivation of olive tree started in the Mediterranean basin (mainly in the regions of Palestine and Syria). Nevertheless, related with the origin of the olive tree there are many different opinions. Archeologists believe that olive trees began being cultivated in the second millennium BC. This idea is supported because of pottery jars and other artifacts which have been found, and in which the population probably stored olive oil. In the early Bronze Age (3,300 - 2,000 BC), it is believed that the population began pruning more efficiently to increase the production of olives. Between 3,200 and 3,100 BC, Palynology (the science which studies pollen) has revealed the presence of *oleaster* pollen in Kopais, Thessaly and Crete (Boskou, 2006). Evidence of the spread of olive pollen throughout Greece corresponding to the second millennium BC has been found. This confirms that, between 2,000 and 1,450 BC, the cultivation of these trees in this region was intense and systematic and played an important role in the economy of the Greek peninsula and its islands.

As can be seen from the aforementioned opinions, the olive cultivar origins are still unclear. In recent years, mathematical algorithms have been used to locate sites of olive cultivar geographic origins. These algorithms consist of multivariable analysis and Bayesian clustering methods. This group of mathematical methods has been shown as one of the most reliable to establish cultivar relationships (Breton *et al.*, 2008; Besnard *et al.*, 2001). Due to the similarities between different varieties of olive trees (166 *oleaster* trees were sampled in eighteen regions of the Mediterranean basin) which were determined by these algorithms and Bayesian clustering methods quantified by probabilities and proportions, the domestication of the olive clearly took place in different regions. The first domestic olive was probably in the Near East, but domestic traces have been found earlier in the west from the Mediterranean basin (Breton *et al.*, 2008). While, studying the changes of olives under domestication on geographical and chronological scales, Terral *et al.* stated that the origins of the cultivated forms created and/or introduced in the north-western regions of the Mediterranean date back to the Bronze Age (Terral *et al.*, 2004).

## 2.1. Expansion of Olive Tree Trade

From Crete, olive oil, cultivations and production techniques were exported to the Greek peninsula, Asia Minor and North Africa, mainly Egypt by Athenian ships (Boskou, 2006; Harwood and Aparicio, 2000). The maritime trade routes throughout the known world established by the Phoenicians made possible the extension of the olive tree trade to Libya, Carthage and the Greek islands (Kapellakis *et al.*, 2008; Boskou, 2006). Then, from Greece, olive tree farming was extended to their Colonies. In fact, the island of Samos was called Elaeophytos which means "planted with olives". In the eighth and seventh centuries BC, the first significant improvement in olive farming occurred due to this cultivation becoming more systematic (Kapellakis *et al.*, 2008). Through their contacts with Greek colonies in Italy, the Romans encountered olive trees, and, its trade was extended by them throughout their empire (Asia Minor, Egypt, *etc.*) (Kapellakis *et al.*, 2008). In this context, olive oil gained importance not only as a staple food, but also as a source of power, raw material for kitchen artifacts, pharmaceutical products, *etc.* Olive tree cultivation expansion continued and in the areas where the olive tree was well-known (primarily Spain, Italy, and Greece) techniques related with its farming and pruning advanced notably. The opposite trend took place in North Africa and other areas taken over by Turks, but was revived later in Arab-controlled countries. By 1,400, Italy had become the greatest producer of olive oil in the world, offering extraordinary quality oil. In the latter part of the fifteenth century, when the Americas was discovered, the exchange of foods and seeds was widely developed. Olive trees were mainly cultivated in regions with similar climatic conditions to those in the Mediterranean area such as Chile and Argentina. In the late 1,700s, Franciscan monks brought the first olive trees to California to the Mission San Diego de Alcalá (Sibbet and Ferguson, 2005). One hundred years later, owing to the demand from Italian and Greek immigrants olive oil imported from Europe, made its commercial debut in the Americas.

The greatest expansion of olive oil production came after the eighteenth century, when large plantings of olive trees, largely relegated to the poorest land, were made to supply the growing populations of the cities. Nevertheless, in the late nineteenth and twentieth centuries, the development of low-cost solvent extraction techniques for seed oils and the use of other sources for light (gas and electricity) resulted in a drop in the demand for olive oil.

Taking into account that the olive tree is able to resist adverse climatic conditions, showing tolerance to saline water and adapting to other types of climatology, olive trees have spread from the Mediterranean basin to other regions with other climatologies such as Australia, South Africa, South America, *etc.* (Sibbett and Ferguson, 2005). Historically, olives tree began to crop after ten years; currently, due to modern technology, although full productivity is not achieved until the seventh or eighth year (in irrigated groves) olive trees begin to crop in the third year, five years earlier than in classical cultivation. This point is a really important improvement in the industrial cultivation of the olive tree increasing its economic return.

## 3. CONCLUSION

The origin of the olive tree has been established by scientific sources as *circa* sixty millenniums ago. Since then, many applications have been found for olive oil such as cosmetic and pharmaceutical products, energy source, food, wood, soap, medicine, *etc.* The expansion of trade routes for foods and other goods throughout the world have made possible that nowadays, extra virgin olive oil is available in nearly every home in the world. In addition, given the improvement of technology related with harvesting, olive trees can be found all over the world.

## REFERENCES

Besnard, G., Breton, C., Baradat, P., Khadari, B. & Berville, A. (2001). Cultivar identification in the olive (Olea europaea L.) based on RAPDS and mitochondrial RFLPs, *Journal of the American Society for Horticultural Science, 126*, 668–675.

Boskou, D. Olive oil: Chemistry and technology. 2nd Edition. Champaign, Illinois: AOCS PRESS; 2006.

Breton, C., Pinatel, C., Medail, F., Bonhomme, F. & Berville, A. (2008). Comparison between classical and Bayesian methods to investigate the history of olive cultivars using SSR-polymorphisms. *Plant Science, 175*, 524–532.

Harwood, J. & Aparicio, R. Handbook of olive oil: Analysis and properties. Gaithersburg, Maryland. Aspen Publishers, Inc; 2000.

Kapellakis, I. E., Tsagarakis, K. P. & Crowther, J. C. (2008). Olive oil history, production and by-product management. *Reviews in Environmental Science and Biotechnology, 7*, 1–26.

Lumaret, R., Ouazzani, N., Michaud, H., Vivier, G., Deguilloux, M. F. & Di Giusto, F. (2004). Allozyme variation of oleaster populations (wild olive tree) (Olea europaea L.) in the Mediterranean Basin. *Heredity, 92*, 343–351.

Sibbett, G. S., Ferguson, L., Coviello, J. L. & Lindstrand, M. Olive production manual. Oakland, California. University of California Agriculture and Natural Resources; 2005.

Terral, J. F., Alonso, N., Capdevila, R. B., Chatti, N., Fabre, L., Fiorentino, G., Marinval, P., Perez Jorda, G., Pradat, B., Rovira, N. & Alibert, P. (2004). Historical biogeography of olive domestication (Olea europaea L.) as revealed by geometrical morphometry applied to biological and archaeological material. *Journal of Biogeography, 31*, 63–77.

Torrecilla, J. S. (2001). Aprovechamiento del alpeorujo. *Agricultura, revista agropecuaria, 832*, 734-737.

*Chapter 2*

# THE LIFE OF THE OLIVE TREE

## ABSTRACT

A description and the main characteristics of the olive tree are described in this Chapter. These topics will help the reader to understand the relationship between biochemistry, agricultural techniques and the growth of the olive tree. In the first part, the genus, influence of roots, trunk, leaves, inflorescence, flowers, fruit, olives and their influence in the growth of the olive tree have been described. The assimilation of chemicals and nutrients and its influence in the cultivation and in olive production is then studied. Finally, the pruning process is defined and its main objectives and methods such as frequency, intensity, differences between un-pruned and pruned olive trees and the best strategies have been shown. Given that the subjects studied here are extensive, a summary of interesting tips and a guide about growing olive trees is provided here.

**Keywords:** History of the olive tree; Olive; Olive tree; Prune; Nutrients assimilation.

## 1. THE BOTANY OF OLIVE TREE

The olive tree from the genus *olea* and species *europaea* produces edible fruits, olives, which belong to the *oleaceae* family. The term of olea europaea L. sativa is used to distinguish it from the wild olive subspecies, *oleaster*. These plants may be seedlings of cultivated varieties spread by birds and other wildlife feeding on the fruit. The cultivated olive tree is diploid. Given that all the olea genera have the same chromosome number ($2n = 2x = 46$), crosses between many of them have been successful (Vossen, 2007; Sibbett *et al.*, 2005). In fact, this specie, the commercial one, it is a hybrid composed of two or more distinct species. The olive flowers contain both male and female parts, but some cultivars are male-sterile and others have only staminate flowers (Doveri *et al.*, 2006)

The olive tree is a long living and evergreen tree. Given that these specimens can overcome adverse climatology, extreme environmental conditions or even mechanical damage, these trees can live for many years. The oldest olive tree in the world is in Kolymbari, Crete (Greece), and is estimated as being 3,000 years old. This ancient tree (its perimeter and radius are greater than 12 and 2 m, respectively) is currently producing olives and is considered as a monument. In addition, there are also other specimens belonging to the time of Jesus, which are more than 2,000 years old. As an example, a "younger" olive tree from Ithaca, Greece that is more than 1,500 years old is shown in figure 1. Although the wood obviously resists the course of the time, their trunks decompose, and therefore, the older olive trees present extremely gnarled trunks, figure 1 (Sibbett *et al.*, 2005).

Figure 1. One of the oldest olive trees in the world, more than 1,500 years old (Ithaca, Greece).

The physical appearance of this type of tree is characteristic; it presents a dense assembly of limbs and the foliage forms a compact structure. In fact, in the case of the wild olive tree, light cannot penetrate to its internal part. Nevertheless, in the well attended olive tree, the pruning process creates void spaces where light can reach the trunks. This is one of the reasons why a pruned olive tree can help vigorous growth, and therefore, its olives can grow throughout the limbs, and then, the production of olives is higher than in the case of a non attended olive tree (Sibbett *et al.*, 2005).

The roots of the olive tree can be classified into two groups viz. those that are short lived (several weeks) and those that develop a secondary enlargement. In general, as with other plants which are suitable for arid or semiarid conditions, the roots of the olive trees penetrate into the ground less than 1.5 m, but they can extend over more than 15 m around the tree. In young trees, the lateral enlargement of their roots is more rapid than those which are mature. In general, the growth of the roots of the olive tree is essential to their optimal performance. In fact, limited extension greatly affects the tree size and its productivity. In addition to their anchorage functions, the roots have three additional functions: (*i*) absorption of water and mineral nutrients, (thanks to the extensive length and very thin diameter brown roots they have a large surface area and are able to selectively absorb nutrients and water from the soil); (*ii*) synthesis of chemicals, in the apical part of the roots produces two important hormones, gibberellins and cytokinins, and abscisic acid which is made in the root cap; (*iii*) storage of carbon (as starch and soluble carbohydrate) and nitrogen (as amino acids and proteins). The latter is absorbed and retained, for later use, mainly in the late summer and autumn (Sibbett *et al.*, 2005).

Figure 2. Both sides of olive tree leaves.

Olive tree leaves are leathery, fibrous, and thick, and their sides have two distinct colors, figure 2. Stomata are only in the lower part of the leaves and these are situated in the peltate trichomes. As a result of these, the olive tree restricts water loss and makes it more drought resistant (Sibbett *et al.*, 2005). In order to protect the olive tree against *bio*tic and a*bio*tic stress, the leaves have many components (waxes, celluloses, carbohydrates, lignins, phenolic substances and oils) in the cuticle, cell wall and cytoplasm of cells (Gucci and Cantini, 2008). These leaves are retained all year round developing fully in less than three weeks and can live more than 3 years. Nevertheless, under stress condition the leaves last less than 2 years. The new and old leaves present different colors and at first sight, the olive tree growth can be detected.

Figure 3. Olive inflorescence prior to opening of all flowers.

The flowering consists of three stages *viz.* transition between vegetative to inflorescence buds which starts in early summer, initiation of flowering begins in November and finally the differentiation of the floral parts. The first two stages depend mainly on two factors: (*i*) the floral development depends on good nutrition mainly relating to the nitrogen concentration (*vide infra*) and water content of the soil; (*ii*) Colder environments have a profound affect on floral development. Inflorescence is produced in the axis of most of the olive tree leaves and, depending on the cultivar, each inflorescence contains between 15 and 30 flowers. At the beginning of each season, the flowers start to grow, there being hundreds of flowers per twig. Olive inflorescence prior to the opening of all flowers is shown in figure 3. All these flowers are small and pale yellow. Olives are *andromonecious,* the same olive tree bears two types of flowers which are present in each season viz. the perfect one containing stamens and pistils (female) and staminate flowers (pollen bearing or male) containing aborted or degenerated pistils and functional stamens. Why the growth of these flowers varies is not well known, it may be dependent on the presence of water, and nutrients during the floral development and/or the excessive population of flowers.

Botanically, the fruit of the *Olea europea* is a drupe and belongs to the same family as almonds, apricots, cherries, damsons, nectarines, peaches, plums, *etc.* (in general all members of the *Prunus* genus) (Sibbett *et al.*, 2005). This fruit consists of pericarp (skin and pulp) and endocarp (kernel pit). The first and the second parts represent 65-83 % and 13 to 30 % of the whole olive, respectively. Depending on the state of ripeness, their color ranges from green to black. Depending on the variety, their weights are between 2 and 20 g. The weight of the olive increases until October-November, then as a consequence of water loss, it starts to decrease. Up to 70 % of the weight of the olive is water. The main composition of olives is shown in table 1. In addition, other constituents such as pectins, organic acids, pigments, phenols, *etc.* can also be found in olives. The oil content, concentrated in the pericarp, begins to appear in August, reaching a maximum in the period from October to December. Although the maturation process of the olive depends on the variety, available water, region, climatologic conditions, *etc.* in nearly all species this process lasts several months (Boskou, 2006). This maturation stage is an important factor in establishing the beginning of the harvest.

**Table 1. Main composition of olives.**

| Compounds | Value (%) |
|---|---|
| Water | 50 |
| Protein | 1.6 |
| Oil | 22 |
| Carbohydrates | 19 |
| Cellulose | 5.8 |
| Minerals (Ash) | 1.5 |

In general, two different types of olives can be produced, those to be used as table olives and those for oil production. The main difference between them consists of the relationship between pulp and kernel of the olives. In the case of olives for oil production, the fruit has a lower pulp to kernel ratio (4/1 to 7/1) in relation to the same ratio of olives for the preparation of table olives (7/1 to 10/1) (Boskou, 2006). In both cases the classification of the product is complex (Torrecilla *et al.*, 2009).

## 2. CHEMICALS ASSIMILATION

As with other type of trees, the growing of the olive needs the presence of chemicals. Some come from the air (carbon in the form of carbon dioxide and oxygen) and others (nitrogen, phosphorus, potassium, *etc.*) are provided by the soil. The growth of the olive tree is based on a group of highly complex assembled *bio*chemical reactions, where inorganic molecules are converted into organic compounds by the photosynthesis process, figure 4. In particular, this consists of the transformation of six molecules of water ($H_2O$) and six molecules of carbon dioxide ($CO_2$) in one molecule of sugar, $[(CH_2O)_n]$ and six molecules of oxygen $[O_2]$ by the activity of chlorophyll (present in the olive tree leaves) which is derived from solar energy. Glucose (sugar in figure 4), made by carbon dioxide taken from the environment, is an important compound in the growth of the olive tree because it can be used to manufacture cellulose, hemicelluloses, starch, pectins, *etc.*, or it can be stored as an energy source for the tree itself or to create new parts of the plant. Nevertheless, during the life of the tree, carbon also can be lost in the crop harvest, pruning and leaf loss and because of diseases (Sibbett *et al.*, 2005). The reverse process of photosynthesis is the respiration process, in which solar energy is not necessary and sugar and oxygen are transformed into water and carbon dioxide, figure 4.

There are five main factors having significant importance in the photosynthesis process and are based on:

- **The penetration capability of light**, that is, solar energy is absorbed by the chlorophyll molecules which are present in the leaves. Only 30 % of the light received is used to carry out the photosynthesis process, and because of this the highest number of leaves possible should be exposed to sunlight. The importance of correct pruning is therefore an essential stage (*vide infra*).
- **Environmental temperature**, as the photosynthesis process is based on *bio*chemical reactions, the temperature at which these reactions take place is vitally important. The optimum temperatures range between 15 and 30 °C.

- **Carbon dioxide concentration**, as this chemical plays a fundamental reactive role, its concentration is important but given that the concentration of carbon dioxide in the atmosphere can be assumed as a constant (0.03 %), this importance is reduced.
- **Nutrient supply**, because the chemicals are used to make one of the most important molecules, chlorophyll.
- **The added water** mainly influences the photosynthesis process in two different ways; firstly, it is fundamental for the reaction. And secondly, it is part of the performance of stomata and limits the carbon dioxide available, figure 4 (Grattan *et al.*, 2006).

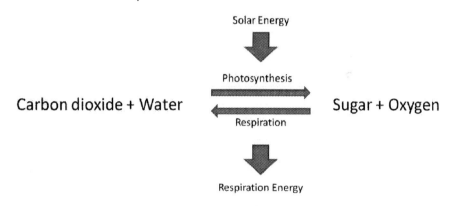

Figure 4. Photosynthesis and respiration processes.

The reverse process of photosynthesis, the respiration process mainly depends on four factors based on:

- **Environmental temperature**, as in other *bio*chemical reactions, the temperature of the reaction affects their kinetics. As in the direct process, the optimum range of the respiration process is between 15 and 30 °C. At lower temperatures the reaction rate decreases, probably due to the change on the fluidity of compounds. At temperatures greater than 30 °C the reaction rate decreases due to the alteration of the enzymes which are required to carry out the *bio*chemical reaction.
- **Oxygen concentration**, this is essential to the respiration process because oxygen is a reagent of the process, but since this is an unlimited reagent, this is not a limiting factor. Nevertheless, given that the roots of olive trees need to have oxygen to grow, water logging can kill olive trees.

- **Solubility of carbohydrates,** sugars are produced in the leaves and these are transported to all parts of the olive tree. These compounds are used in the respiration process or stored as a food reserve. This storage will affect the posterior photosynthesis process.
- **Physiologic factors.** All factors in this group related with the *bio*chemical reaction (respiration process) will be taken into account, e.g. leaf age or growth cycles, *etc.*

The *bio*chemistry and physics of the photosynthesis process is not the aim of this book and (see Gucci, R. & Cantini, 2008; Boskou, 2006 e.g.), this process will not be described in more detail here.

## 2.1 Nitrogen Cycle in Orchards and its Assimilation

Although, nitrogen ($N_2$) is one of the most abundant chemicals in the Earth's atmosphere (78 %), this molecule is not fully suitable for use by plants. Although the olive tree takes some nitrogen from the atmosphere by photochemical reactions and lightning discharges, this compound is mainly taken from soil. It is then converted from inorganic into organic molecules in a process known as the nitrogen fixation process (*vide infra*). In the first stage, the atmospheric nitrogen is converted into proteinaceous material and ammonium by means of microorganisms and these new compounds are retained in the soil. The nitrogen is then reabsorbed by the plants and finally returns to the soil as residues. The whole process is known as the nitrogen cycle. The storage of nitrogen in the soil is carried out by the mineralization process. This process consists of the formation of amino ($NH_2$) through enzymatic hydrolysis of proteins. These amino compounds are used by the microorganisms in order to produce ammonia ($NH_4^+$) by ammonification. The latter can be absorbed by plants and interchanged with other ions which are in the soil. The nitrosomas are able to convert the ammonia into nitrite ($NO_2^-$) and then it is converted by the nitrobacteria group in nitrate ($NO_3^-$). This latter nitrogen form and ammonia are used by the olive tree to produce proteins, and when the leaves fall, these return to the soil. The olive tree loses nitrogen by the denitrification, runoff, volatilization, pruning, leaching to groundwater, and crop removal processes. These losses are highest when the soil has a low oxygen content and a higher nitrate concentration, and therefore, requires to be fertilized to maintain an adequate nitrogen concentration (Sibbett *et al.*, 2005). Nevertheless, the different nitrogen forms are assimilated by the olive tree depending on physiological age, light exposure, tree age, *etc.*

# 3. PRUNING OLIVE TREES

Pruning trees is not only the process of removing certain above-ground elements from a plant, but is also one of the most critical practices to achieve successful orchards and is one of the most powerful tools for manipulating plant growth and reproductive processes (Gucci and Cantini, 2008). To carry out this process adequately, many physiologic and biologic aspects must be kept in mind. Therefore, pruning techniques vary with the species, with particular emphasis on the age of the olive tree. The pruning of young and mature olive trees is different (Gucci and Cantini, 2008; Sibbet, 2005). The response to the pruning of the trees varies with the cultivar, tree age, vigor, crop load, history of the plant, *etc.* A more detailed explanation about this can be found elsewhere (Lambers *et al.*, 1998; Gucci and Cantini, 2008; Sibbet, 2005).

The stored carbohydrates and the potential of the leaf surface to grow are eliminated by the pruning process. Comparing pruned with un-pruned olive trees, different effects should be highlighted: (*i*) in the first case, as a consequence of the reduction in leaves, the growth of the roots, and therefore the nutrients, are reduced and the competition for these nutrients and carbohydrates is also reduced at the growing stage; (*ii*) in the case of un-pruned trees, the leaf areas are greater; (*iii*) the pruning process induces an increase in the available nitrogen. When small diameter twigs are cut in a regular overall manner, the metabolism and growth of the olive tree are stimulated. Nevertheless, to prevent lateral shoots, cutting the apex branches is required (Gucci and Cantini, 2008; Sibbet, 2005).

In young olive trees, pruning is required mainly to the assist formation of a well balanced canopy structure. With this structure the young tree can absorb the maximum amount of light possible in the totality of the tree, favoring air circulation, and once the tree is in full production, support the crop load and make the mechanical harvesting easier. In the mature olive tree, its pruning is also appropriate to favor light penetration and air circulation, renew the fruiting surface to achieve high yields, maintain vegetative growth and skeleton structure, prevent the ageing of the canopy, making it easy to carry out an adequate control of diseases and pests, minimizing alternate bearing (the olive tree is characterized by alternative heavy and light olive production), limiting tree space, removing frost damage, *etc.* (Gucci and Cantini, 2008). To summarize, the principal objectives of pruning for both young and mature olive trees are shown in table 2. The pruning is governed by the final destination of olive tree, e.g. production of table olives (more importance should be given to promote light penetration) or olive oil, ornamental applications (increasing the aesthetic values), *etc.* (Sibbet, 2005; Gucci and Cantini, 2008).

**Table 2. Objectives for pruning depending on the age of the olive trees.**

| Younger olive tree | Mature olive tree |
|---|---|
| Allow earlier olive production<br>Form tree canopy structure | Renew fruit shoots.<br>Prevent ageing of the canopy.<br>Balance vegetative and reproductive activities.<br>Eliminate dead wood.<br>Managing alternate bearing.<br>Repair damage to the canopy after stress.<br>Rejuvenate old or abandoned trees. |
| General objectives | |
| Maximize sunlight exposure.<br>Maintain tree canopy structure.<br>Limit tree space.<br>Favor air circulation.<br>Control tree size.<br>Achieve high yields.<br>Improve aesthetic value of trees.<br>Improve fruit quality. | |

In order to achieve higher productivity yields and prevent risk of infection, attack by insects, *etc.,* the mature olive tree should be pruned between spring and summer, depending on the region. Although pruning in summer is used to stimulate the lateral growth of the young olive tree and to improve the quality of the table olives, it is not common in those plants dedicated to olive oil. In mild climates, the pruning process can start in winter, but in colder climates, because of potential frost damage, starting the pruning process earlier is a risk. In all cases, pruning should not take place after full bloom. In general, at least one pruning process per year is recommendable. To recap, taking into account that the pruning timing affects the response of the plant, the cultivator should consider economic aspects to avoid excessive pruning costs (Sibbet, 2005; Gucci and Cantini, 2008).

In general, the intensity of pruning depends on age, cultivar, crop load, soil fertility, water, *etc.* From the point of view of age, severe pruning is usually applied to mature olives tree and a less rigorous process should be applied to younger trees or when these plants are being cultivated in irrigated fertilized soil. In arid soil cultivation, more aggressive pruning is required in order to limit leaf growth and so reduce water consumption. The intensity of pruning can be respectively classified as light, moderate or severe if 17, 25 – 30 % or 50 % of the wood is removed. Nevertheless, the intensity can be calculated as a function of the number of sprouts and the average length of branch growth in one year in the mean part of canopy. The general strategy is based on pruning as little as possible, reducing the frequency of pruning, adapting an open canopy to make light penetration and air circulation easier and using irrigation and fertilization to stimulate olive growth. As a minimum, in ideal circumstances, the olive tree should be pruned once every 8 or 10 years. In the final analysis, the decision regarding pruning is based on economic factors.

## 4. CONCLUSION

A description of the life of olive trees and the characteristics which most influence their growth, such as roots, leaves, flowers or olives are described here. Why the process of nutrient assimilation, the nitrogen presence in soils, environmental temperature, or the available water are really important in the life of olive trees in order to obtain the maximum crop has been explained. Also, the pruning process has been summarized here, given general working ranges and tips. These topics help the reader to understand the relation between *bio*chemistry, agricultural culture and the growth of the olive tree.

## REFERENCES

Boskou, D. Olive oil: Chemistry and technology. 2nd Edition. Champaign, Illinois: AOCS PRESS; 2006.

Doveri, S., O'sullivan, D. M. & Lee, D. (2006). Non-concordance between Genetic Profiles of Olive Oil and Fruit: a Cautionary Note to the Use of DNA Markers for Provenance Testing. *Journal of Agricultural and Food Chemistry. 54*, 9221-9226

Grattan, S. R., Berenguer, M. J., Connell, J. H., Polito, V. S. & Vossen, P. M. (2006). Olive oil production as influenced by different quantities of applied water. *Agricultural water management, 85*, 133 – 140.

Gucci, R. & Cantini, C. Prunning and training systems for modern olive growing. Collingwood, Australia. CSIRO publishing; 2008.

Lambers, A. F., Chapin, F. S. & Pons, T. L. Plant physiological ecology. Springer Verlag, New York. 1998.

Sibbett, G. S., Ferguson, L., Coviello, J. L. & Lindstrand, M. Olive production manual. Oakland, California. University of California Agriculture and Natural Resources; 2005.

Torrecilla, J. S., Rojo, E., Oliet, M., Domínguez, J. C. & Rodríguez, F. (2009). Self-organizing maps and learning vector quantization networks as tools to identify vegetable oils. *Journal of Agricultural and Food Chemistry, 57*, 2763–2769.

Vossen, P. (2007). Olive Oil: History, Production, and Characteristics of the World's Classic Oils. *Hortscience, 42*, 1093-1100.

# SECTION 2. TABLE OLIVES

*Chapter 3*

# TABLE OLIVES

## ABSTRACT

Table olives come directly from the fruit of the olive tree which is a small drupe 1–2.5 cm long. Although these fruits are mainly produced in the European Community (Spain is the first producing and exporting country), in the last half century, as a result of substantially increased demand in table olives, their production has widened throughout the world (California, Australia, China, *etc.*). These fruits can be classified depending on their color, treatment applied, combination of both, *etc.* In addition, in order to be sold, depending on their defects, every type should be classified into three different categories (extra, first and second categories). Due to the fact that table olives present a high concentration of *glycoside oleuropein*, an unpleasant and hard bitter taste is present, although this is not detrimental to health, the olives cannot be eaten directly and need to be physicochemically treated. In this Chapter, a brief history of table olives, their trade activities, their main classifications and description of the elaboration processes of green, natural black and black ripe olives can be found. As all relevant aspects cannot be studied in detail here, scientific references are provided.

**Keywords:** Green olives; Black olives; Black ripe olives; Elaboration of table olives; Brining and fermentation; Lye treatment.

## 1. TABLE OLIVES AND TRADE ACTIVITIES

The Olive tree from the genus *olea* and species *europea* belongs to the family *oleaceae*. Its fruit, the olive, is a small drupe 1–2.5 cm long, thinly-fleshed and smaller in wild plants than in orchard cultivars (see Chapter 2). A classical description of olives included in Koehler's Medicinal-Plants which was published in the late nineteenth century is shown in Figure 1.

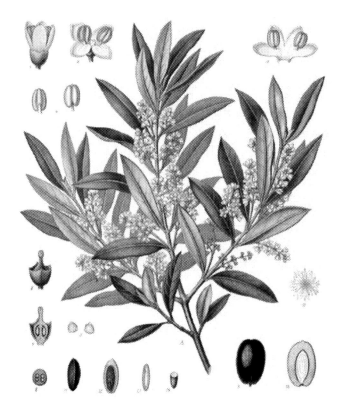

Figure 1. Classical olive description from Koehler's Medicinal-Plants 1887.

Table olives are harvested at the green stage or left to ripen to a rich purple color (black olive). They have a low sugar content (2.6-6 %) compared with other drupes (12 % or more) and a high oil content (12-30 %) depending on the time of year and variety. Although the table olives pose no damage to health, without treatment these fruits present a strong bitter taste which is due to the presence of oleuropein [(4S,5E,6S)-4-[2-[2-(3,4-dihydroxyphenyl)ethoxy]-2-oxoethyl]- 5-ethylidene-6-[[(2S,3R,4S,5S,6R)-3,4,5-trihydroxy-6-(hydroxymethyl)- 2-tetrahydropyranyl]oxy]-4H-pyran-3-carboxylic acid, methyl ester] (distinctive of the olive, *vide infra*).

## 1.1. Table Olives in the Past

The origin of table olives and olive trees are joined. Table olives came from *olea oleaster* commonly called the wild olive tree. It is native to the Mediterranean basin, central Asia and some parts of Africa (see Chapter 1). These trees produce edible fruits and have short branches, round to oval leaves and small round fruits with large stones and little flesh. Currently, these types of trees can be found especially in Greece and Italy. This type of tree is the progenitor of olive tree currently cultivated, called the domestic olive tree *olea europaea L.* (Kailis and Harris, 2007). Some authors state that olive trees were "domesticated" in Crete around 2,500 BC and also in Iran and Mesopotamia. From the latter, the olive was spread throughout the world by different empires and cultures. Initially, through the south and west (Palestine and Anatolia), then, by the trading activities of Greeks and Phoenicians, the olive spread to the whole Mediterranean basin (see Chapter 1). Although olives are not indigenous outside these regions, from these geographical points, the olive tree has extended through time to south (Chile, Argentina) and north America (Mexico, California), Australia, South Africa, New Zealand, China, India, Japan, *etc.* (Lumaret *et al.*, 2004; Boskou, 2006; Kailis and Harris, 2007). Currently, there are more than five natural subspecies distributed all over the world *viz. olea europaea europaea* (Europe), *olea europaea cuspidata* (Eritrea, Iran, China, *etc.*), *olea europaea guanchica* (Canaries), *olea europaea maroccana* (Morocco), *olea europaea laperrinei* (Algeria, Sudan, Niger and India), *etc.*

# 2. TABLE OLIVES PRODUCTION PROCEDURE

In this section some groups of techniques to elaborate edible table olives are described. The first methods used and their influence on current techniques are summarized. Then, table olive types and trade categories are briefly described. Finally, the current techniques used to produce edible table olives are described.

## 2.1. First Techniques to Produce Table Olives

Although olives have been an important foodstuff and essential food in the diet of the inhabitants of the Mediterranean basin and the Middle East, when the first table olive was eaten is unknown. It can be assumed that the first table olive was eaten directly from the tree or the ground. Due to their bitter taste, processing of the edible table olive was required. Probably, through trial and error, the inhabitants and farmers developed an adequate procedure to produce edible olives. Then, this knowledge was handed down with improvements being made from generation to generation.

Given that table olives came mainly from the Mediterranean basin, it is understandable that one of the processing stages to eliminate their undesirable bitter taste consists of immersing the table olives in saline water (seawater) and/or drying them in the sun. In addition, to storing table olives in salt water for long periods, the olives were flavored with herbs gathered from hillsides to improve their sensory characteristics. In the middle of the ninetieth century, alkaline wood was used to reduce the bitterness of the olives. These compounds are the precursor of the modern lye treatment. In the last 100 years, table olive processing has passed from small to large scale production. This scaling up is being developed mainly in countries with great hopes of substantially increasing their involvement in the olive sector such as Morocco, Turkey, Argentina, Australia, *etc.* (Kailis and Harris, 2007).

## 2.2. Types of Table Olives

The United Nations Conference for the Negotiation of a Successor Agreement to the International Agreement on Olive Oil and Table Olives, 1986, as Amended and Extended, 1993, having met in Geneva from 25 to 29 April 2005 and updated in 2009, among other subjects, have defined the meaning of table olives as (United Nations Conference on Trade and Development, 2009)

*"The product prepared from the sound fruits of varieties of the cultivated olive tree which are chosen for their production of olives particularly suited to curing, and which are suitably treated or processed and offered for trade and for final consumption."*

As a consequence of the large number of table olive types, different types of classification have been proposed. According to the surface color of the table olive or type of olives, intergovernmental organizations (African, Caribbean and Pacific Group of States, European Commission, European Council and International Olive Oil Council) has classified table olives into one of the three following types (United Nations Conference on Trade and Development, 2009). An example of this classification is shown in Figure 2:

Figure 2. Table olives; (a) green olive (Manzanilla variety); (b) Olive turning color (Black Cacereña variety); (c) Black olive (Cacereña variety).

- **Green olives**: fruits harvested during the ripening period, prior to coloring and when they have reached normal size. Their color may vary from clear green to straw yellow. These olives are firm, sound, resistant to slight finger pressure, and without marks other than the natural pigmentation.

- **Olives turning color**: fruits harvested before the stage of complete ripeness is attained, at color change. Their colors are rose, wine rose or brown.
- **Black olives**: fruits harvested when fully ripe or slightly before full ripeness is achieved. Their color varies according to the production region and time of harvesting. They may vary in color from reddish black to violet black, deep violet, greenish black or deep chestnut.

According to the chemical treatments or trade preparations of the table olive, the International Olive Oil Council has created the aforementioned classification of table olives into one of the five following types (Codex Alimentarius Commission, 2004; International Olive Oil Council. 2004; Sánchez *et al.*, 2006):

- **Treated olives:** "Green olives, olives turning color or black olives that have undergone alkaline treatment, then packed in brine in which they undergo complete or partial fermentation, and preserved or not by the addition of acidifying agents: (a) Treated green olives in brine; (b) Treated olives turning color in brine; (c) Treated black olives."
- **Natural olives:** "Green olives, olives turning color or black olives placed directly in brine in which they undergo complete or partial fermentation, preserved or not by the addition of acidifying agents: (a) Natural green olives; (b) Natural olives turning color; (c) Natural black olives."
- **Dehydrated and/or shriveled olives:** "Green olives, olives turning color or black olives that have undergone, or not, mild alkaline treatment, preserved in brine or partially dehydrated in dry salt and/or by heating or by any other technological process: (a) Dehydrated and/or shriveled green olives; (b) Dehydrated and/or shriveled olives turning color; (c) Dehydrated and/or shriveled black olives."
- **Olives darkened by oxidation:** "Green olives or olives turning color preserved in brine, fermented or not, darkened by oxidation in an alkaline medium and preserved in hermetically sealed containers subjected to heat sterilization; they shall be a uniform black color."
- **Specialties:** "Olives may be prepared by means distinct from, or additional to, those set forth above. Such specialties retain the name "olive" as long as the fruit used complies with the general definitions laid down in this Standard. The names used for these specialties shall be sufficiently explicit to prevent any confusion, in purchasers' or

consumers' minds, as to the origin and nature of the products and, in particular, with respect to the designations laid down in this Standard."

Another type of classification has been based on the style of olives *viz.* whole olives, stoned (pitted) olives, stuffed olives, salad olives, olives with capers, olive paste, *etc.* (Codex Alimentarius Commission, 2004). Whereas, the Spanish Regulation "Real Decreto 1230/2001". Reglamentación Técnico Sanitaria para la elaboración, circulación y venta de las aceitunas de mesa" (Boletín Oficial del Estado, 2001; Sánchez *et al.*, 2006) distinguishes four elaboration types according to surface color: Green, turning color, natural black and ripe olives.

In addition, according to the percentage of defects, every aforementioned type of olive can be also classified into three categories (Codex Alimentarius Commission, 2004):

- **Extra** or Fancy: "The high quality olives endowed to the maximum extent with the characteristics specific to the variety and trade preparation are considered as belonging to this category. Notwithstanding, and providing this does not affect the overall favorable aspect or organoleptic characteristics of each fruit, they may have very slight color, shape, flesh-firmness or skin defects. Whole, split, stoned (pitted) and stuffed olives of the best varieties may be classified in this category, providing their size exceeds 351/380".
- **First** choice or Select: "This category covers good quality olives with a suitable degree of ripeness and endowed with the characteristics specific to the variety and trade preparation. Providing this does not affect the overall favorable aspect or individual organoleptic characteristics of each fruit, they may have slight color, shape, skin or flesh-firmness defects. All types, preparations and styles of table olives may be classified in this category, except for chopped or broken olives and olive pastes."
- **Second**, or Standard: Although they cannot be included in the two previous categories, this category comprises good quality olives which comply with the essential composition required to be fit for consumption.

These defects are classified aesthetically, marks on the skin of the olives, damage to the mesocarp, stems attached to olives, abnormal texture and color, *etc.* The main trade characteristics for green olives, olives darkened by oxidation, olives turning color and black olives are shown in table 1 (Codex Alimentarius Commission, 2004).

**Table 1. Maximum tolerances of the trade categories shown as a percentage (Codex Alimentarius Commission, 2004).**

| | Extra category | | | First category | | | Second category | | |
|---|---|---|---|---|---|---|---|---|---|
| | GO | ODO | OTC+BO | GO | ODO | OTC+BO | GO | ODO | OTC+BO |
| **Stoned (pitted) or stuffed olives** | | | | | | | | | |
| Stones and/or stone fragments | 1 | 1 | 2 | 1 | 1 | 2 | 1 | 1 | 2 |
| Broken fruit | 3 | 3 | 3 | 5 | 5 | 5 | 7 | 7 | 7 |
| Defective stuffing placed-packed | 1 | 1 | 1 | 2 | 2 | 2 | - | - | - |
| Defective stuffing random-packed | 3 | 3 | 3 | 5 | 5 | 5 | 7 | 7 | 7 |
| **Whole olives, stoned or stuffed** | | | | | | | | | |
| Blemished fruit | 4 | 4 | 6 | 6 | 6 | 8 | 10 | 6 | 12 |
| Mutilated fruit | 2 | 2 | 3 | 4 | 4 | 6 | 8 | 8 | 10 |
| Shrivelled fruit | 2 | 2 | 4 | 3 | 3 | 6 | 6 | 6 | 10 |
| Abnormal texture | 4 | 4 | 6 | 6 | 6 | 8 | 10 | 10 | 12 |
| Abnormal Color | 4 | 4 | 6 | 6 | 6 | 8 | 10 | 10 | 12 |
| Stems | 3 | 3 | 3 | 5 | 5 | 5 | 6 | 6 | 6 |
| Cumulative maximum of tolerances for these defects | 12 | 12 | 12 | 17 | 17 | 17 | 22 | 22 | 22 |
| Harmless extraneous material (unit per kg or fraction) | 1 | 1 | 1 | 1 | 1 | 1 | 1 | 1 | 1 |

GO       Green olives
ODO     Olives darkened by oxidation
OTC     Olives turning color
BA       Black olives

## 2.3. Modern Techniques to Produce Table Olives

Depending on the processing type and the family of the olive tree, olives should be picked at the appropriate moment of ripeness, i.e., green, turning color or black. As *glucoside oleuropein* is in olives and these chemicals are responsible of their unpleasant bitterness, these fruits are not edible in this state. They require to be chemically treated in order to make them edible, i.e., to remove this glucoside and render the olive edible, the fruit must be cured (*vide infra*).

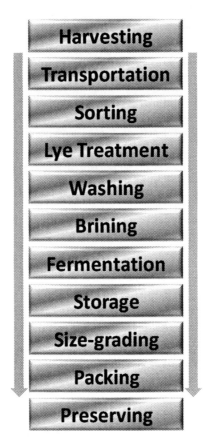

Figure 3. Manufacture process of Green olives.

Once table olives have been harvested, their preparation is influenced by local customs. For instance, in Spain Green Sevillian style olives are popular, whereas, in Greece, the consumer prefers naturally black, ripened in brine, or salt dried olives. In Italy, the preference is for sun dried or heat dried olives. Industrially, depending on the type of table olive, slightly different treatments should be applied to convert these fruits into edible table olives. The most common procedures to treat green olives have been described, and its principal modifications to manufacture black and black ripe olives have been also summarized here. In most cases, the manufacturing process consists of three stages: firstly, the table olives are harvested and transported to the factory. In the factory, lye treatment, washing and brining treatments are applied. Finally, after the fermentation and storage period, the table olives are packed and in some cases pasteurized. A diagram of this manufacture process is given in figure 3 and a description in more detail is given below.

## Harvesting, Transporting and Storage of Table Olives

Green olives should be harvested when the fruit reaches its maximum size and prior to their color becoming pale yellow. Regarding natural black olives, these should be picked when they are completely ripe. In the case of black olives, although currently, they are picked when showing a green color, these should be harvested when these have a yellow-straw tone. As in the case of virgin olive oil, while manual picking is more expensive than mechanical harvesting, in order to limit damage, the fruits should ideally be picked by hand (see Chapters 4 and 5).

In order to cause minimum damage to the olives, and decrease the proportion of damaged olives, the transportation process should be carried out using padded bags (no more than 20 kg) or containers (no more than 500 kg). In order to maintain the quality of table olives recently picked and to prevent peeling of olives during the following process which removes the bitterness from the olives, some authors have proposed transporting olives in dilute sodium hydroxide solution ranging between 0.3 and 0.4 % w/v (Vega et al., 2005; Sánchez et al., 2006).

During the storage process, the process of respiration of the stored fruits appears. Because of this, in one hour, a kilogram of black ripe olives at 5 °C generates up to 20 mL of carbon dioxide increasing the storage temperature. As a consequence of this generation, heat flow (around 4,640 kJ $t^{-1}$), the deterioration of the olive is accelerated. To prevent the respiration heat accumulation in any area of the storeroom, an effective air flow through the olives is required. In addition, during a period of one hour, a kilogram of ripe black olives or mature green olives at 20 °C generates up to 0.5 mL or 0.1 mL of ethylene, respectively. The amount of ethylene produced increases with the environmental temperature. As this gas can accelerate the loss of green and softening of the mature green olives, the concentration of this compound in the storeroom should be controlled (Sibbet et al., 2005). In short, desirable commercial and organoleptic characteristics can be lost during the storage of table olives.

To reduce or even eliminate these defects, sorting fresh table olives and removing immature and overripe culls would be the first step in successful storage (Sibbet et al., 2005). In addition, depending on the maturity, variety and cultivar of the picked olives and the chemical composition of table olives (Agar et al., 1998), to inhibit the undesirable changes and degradation reactions, different alternative storage systems have been proposed. These consist of storage space under a controlled atmosphere, where the user can decide its environmental condition (temperature, relative humidity, *etc.*) and the time of storage of the table olives. Nanos *et al.* affirmed that a variety of green olive (*conservolea*) can be stored up to 37 days at 5 °C in air or for up to 22 days at 7.5 °C and 2 kPa $O_2$ plus 5 kPa carbon dioxide without appreciable change to the taste characteristics (Nanos *et al.*, 2002). Whereas, the storage of Green Manzanillo olives at temperatures lower than 5 °C causes serious damage (Agar *et al.*, 1998). As Arroyo-López and collaborators stated, one of the best conditions to store green table olives consists of treating them with a combination of two washing waters containing sodium chloride (5 %) and sodium metabisulfite (0.1 %) or saturated carbon dioxide, followed by immersion of the fruits in sodium chloride brine (15 %) with sodium metabisulfite (0.1 %) or brine in conjunction with saturated carbon dioxide added (Arroyo-López *et al.*, 2007).

## *Lye Treatment, Washing and Brining of Table Olives*

An essential stage in the curing of table olives is the lye treatment with sodium hydroxide ranging 1.5 and 3 % (Maldonado *et al.*, 2008). Its main objective is the elimination of the bitter taste of the fresh fruits. This objective is achieved by chemical hydrolysis of *glycoside oleuropein*, which is mainly responsible for this characteristic and is the main phenolic compound in the flesh (Medina *et al.*, 2007; Sibbett *et al.*, 2005). In some varieties, to avoid the peeling process, a previous resting period of 1 or 2 days is required. This resting period can be partially substituted by the aforementioned transportation of fruit in dilute solutions of sodium hydroxide (*vide supra*). For a given operational condition and cultivar type, the concentration of lye should be adjusted. Currently, to avoid the peeling effect and make the penetration of sodium hydroxide easier, this treatment is being carried out at 18 °C (Sánchez *et al.*, 2006). Higher temperatures and concentrations of sodium hydroxide led to an increase in the permeability of the skin of the table olive. As lye is an environmentally hazardous material, this solution can be re-used taking care to maintain the sodium hydroxide concentration constant. According to experiments carried out at both a pilot plant and on an industrial scale, the alkali solution can be re-used up to 15 and 7 times, respectively.

Once the lye treatment has been completed, the solution is removed and to eliminate the excess of alkali solution, the table olives are washed with tap water at least once. This stage can last up to 15 hours. Depending on the variety, an excessive washing period is not advisable because essential compounds required in posterior stages, can be dissolved in water and lost. In arid lands, the washing water can be re-used after the application of appropriate treatment.

Once the washing stage is over, the water must be removed. Then, the table olives are input into a solution composed of brine (10–11 % w/v). Ammonium nitrate brines can also be used for periods longer than 15 weeks without spoilage. In this case, during the preparation for canning, the elimination is more difficult (Sibbett *et al.*, 2005). Due to the alkali that is released by the fruits, at the beginning of the brining process the pH is around 10 units. During this stage, fermentation and storage are carried out.

## *Fermentation and Storage Stages*

At the beginning of this stage, due to the fact that acid compounds begin to be generated, pH decreases from 10 to around 4 units. The reason for this decrease is the growth of lactic acid bacteria. Commonly, the fermentation process starts spontaneously. Nevertheless, the inoculation of compounds such as *Lactobacillus pentosus* 5138 or *Enterococcus casseliflavus* and *Lactobacillus pentosus* CECT5138 initiates and accelerates this process (de Castro *et al.*, 2002). As a consequence of the activity of these bacteria, during the two first days of brining, the acidity of the solution increases with the consumption of fermentable sugar. During fermentation, the production of elenolic acid and glucose is produced by hydrolysis of elenolic acid glucoside. The compounds are used by the microorganisms present in the brine to maintain the microbial activity for a longer period of time. The physicochemical characteristics of the fermentation process and the compound generated depend mainly on the olive variety and process. Montaño *et al.* studied more than 160 fermented brines from green olives of the Manzanilla, Hojiblanca and Gordal varieties in five companies and in two consecutive seasons. They found that the Hojiblanca olives present values of pH, combined acidity, and volatile acidity significantly higher than those in the Manzanilla and Gordal varieties, reflecting different processing conditions. Nevertheless, the volatile/total acidity ratio between varieties or seasons did not differ greatly (Montaño *et al.*, 2003).

When the sugars have been consumed, the fermentation process has finished, and the storage period commences. During this latter stage, as a consequence of the *Propionibacterium* bacterium growth, acetic and propionic acids are produced by the lactic acid made previously. Longer storage periods spoil the fruits. But this degradation process can be reduced or even eliminated by increasing the concentration of sodium chloride up to 9 % (w/v) at the end of the main lactic fermentation process (Sánchez *et al.*, 2006).

Natural black olives are not treated with alkali and therefore the fermentation process is slower than in the aforementioned case. Traditionally, the olives are placed in a brine solution with a sodium chloride concentration ranging from 6 % in colder regions to 10 % depending on the environmental temperature. The elimination of the bitter taste can only be carried out by the solubilisation of *glycoside oleuropein* in brine. The fermentation process can be carried out in aerobic or anaerobic conditions and in both cases the operational conditions of this process are influenced by the initial pH value and the sodium chloride concentration. In the aerobic case, to remove the carbon dioxide produced during the fruit respiration, from 10 to 30 % by volume of air (in relation to the fermented volumes in every hour) is introduced in the fermentation vessel per hour (Sánchez *et al.*, 2006). The four main advantages of carrying out the fermentation process via the aerobic mode in comparison with the anaerobic one are (*i*) lower incidence of gas pocket spoilage; (*ii*) the shriveling of olives is eliminated; (*iii*) the fermentation process takes less time (three months versus 8-10 months in the anaerobic case) because the air bubbles improve the contact between reactants and (*iv*) as a consequence of the different polymerization of anthocyanin, the flavor, texture and color is improved (Romero *et al.*, 2004; Sánchez *et al.*, 2006).

Concerning black (ripe) olives, the fermentation and storage stages are carried out in a similar way as with natural black olives. Throughout the storage stage, the sodium chloride concentration should be increased from 4 to 9 %. With this type of olive, this stage can be carried out in different ways depending on the region. Commonly, in the Unites States of America, the fermentation process is carried out by anaerobic methods in aqueous solutions acidulated by acetic and lactic acid and without salt. In Spain, the olives are conserved in an acid medium (Sánchez *et al.*, 2006). A description in more detail can be found elsewhere (López-López *et al.*, 2009; Sánchez *et al.*, 2006; Benítez *et al.*, 2001).

### *Packing and Pasteurization*

Once the fermentation process is completed and all other characteristics comply with the specifications required prior to packing and the posterior consumption of the table olives, three complementary operations are required to increase their appeal to the consumer. Firstly, faulty olives are removed. And then, the fruits are classified according to size. And finally, the fruits are grouped and stored in plastic drums until being packed. This last operation is useful to reduce the variability related to physical (size) and even the classification of the olives from a chemical point of view.

The final stage of the production process is to assure the physicochemical stability of the table olives throughout their commercial life. In the past, to prevent the growth of microorganisms, table olives were sold in aqueous solutions composed of sodium chloride which presents high values of free acidity (pH less than 3.5). The most common technique used to stabilize table olives consists of packing the fruits in fresh brine. An interesting alternative to save water consists of re-using the fermentation brine. But trying to satisfy consumers, the application of sodium chloride has been progressively reduced and other methods are emerging. One of these techniques is based on heating the fruit during a short period of time, and is called pasteurization. The minimum value of thermal lethality is fixed to 15 $UP^{5.25}_{62.4°C}$ (15 minutes at 62.4 °C) to guarantee the proper preservation of the table olive (International Olive Oil Council, 2005). Due to the high thermal resistance of propionic bacteria, this microorganism is taken as a reference of the process. To preserve the table olives unaltered and maintain a constant quality level throughout their commercial life, during the pasteurization process the table olives are heated at about 80 °C. Although it is not customary, hot brine (around 80 °C) can be used to prevent sedimentation from bacterial growth, (Sibbett et al., 2005).

When the olives are packed in plastic pouches which cannot be pasteurized because the heat would accelerate the polymerization of ortho-diphenols, other preservation methods are required such as packing in a modified atmosphere (vacuum, addition of an inert gas, etc.), addition of authorized preservatives, maintaining the fruit at refrigerator temperature, etc. The containers should be made of material that ensures the correct preservation of the fruits. Habitually, table olives are sold in containers made of metal, tin, glass, plastic or any other material except wood.

The conservation of natural black olives can be carried out by the thermal pasteurization of the packed product in similar condition to green olives (Sánchez et al., 2006; Codex Alimentarius Commission, 2004). In relation to ripe olives, these can be preserved by heat sterilization which destroys spores of *Clostridium botulinum*. To ensure proper conservation during its commercial life, a minimum cumulative sterility value equal to 15 $F_o{}^{10}_{121°C}$ (15 minutes at 121 °C) should be reached (Codex Alimentarius Commission, 2004). Also, these can be in acidified solutions at pH less than 4.5 (in lactic or gluconic acid) in plastic pouches (Sánchez et al., 2006).

## 2.4. Commercialization of Table Olives

About 10 % of the total olives harvested are used to produce table olives. As can be seen in figure 4, in the last two decades, in line with the increase table olive consumption, its production has doubled. The International Olive Oil Council explains that table olive consumption has increased for four main reasons, *viz.* purchasing power of consumers, better presentation of table olives, population increase and the continuing improvement in table olive quality. As with the virgin olive oil trade, as a consequence of the notable increase of production of table olives in non European Community countries, it is hoped that during the 2008/09 season table olive production will increase to nearly 1.4 million tones. In particular, apart from European Community (6.94 $10^5$ t), the countries with the highest production of table olives in the world are Egypt, Morocco, Syria and Turkey with the respective production of 4.0 $10^5$, 1.1 $10^5$, 1 $10^5$ and 2.5 $10^5$ t of table olives. In the European community (using the estimations calculated by International Olive Oil Council for 2008/09 season), Spain, Greece and Italy are the regions with the highest productivity with 4.75 $10^5$, 1.2 $10^5$ and 0.8 $10^5$ t of table olives, respectively (International Olive Oil Council, 2009). The countries that import and export the largest quantities of table olive in the world are shown in table 2. During the 2008/09 seasons, it is estimated that the European Community will export around 38 % of its production and 13 % of the total export of table olives in the world. Whereas, in the Unites States of America, the consumption of table olive is mainly imported (about 66 %) (International Olive Oil Council, 2009).

Due to the innumerable types of table olives available to consumers in the world, the most representative varieties from the most productive country (Spain) will be shown here. In Spain, the most representative and extensive varieties of table olives are Manzanilla (Caceres, Salamanca, Badajoz and Seville), Hojiblanca (Cordoba, Malaga, Seville and Granada), Carrasqueña (Badajoz), Cacereña (Caceres and Salamanca) and Gordal (Seville). Their production values represent respectively 35, 35, 9, 8 and 5 % of the total table olive production in Spain and the average of yearly table olive consumption is about 3.4 kg of table olives per capita (Ministerio de Agricultura, Pesca y Alimentación, 2009). Table olives belonging to the Manzanilla and Cacereña varieties are shown in figure 2. The main composition of the Manzanilla, Hojiblanca and Gordal varieties are shown in table 3. Every season, these species are harvested between September and November.

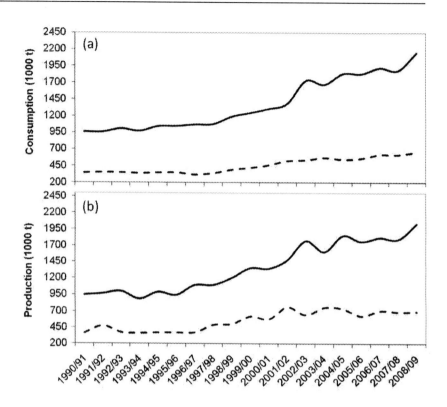

Figure 4. Consumption (a) and production (b) of table olives in the world (solid line) and in the Europe Community (dotted line). Production and consumption during the 2008/09 seasons have been estimated by the International Olive Oil Council. (International Olive Oil Council, 2009)

**Table 2. Estimated amount of imports and exports of table olives in the world during the 2008/09 season (International Olive Oil Council, 2009).**

| Exports (1,000 tonnes) | | Imports (1,000 tonnes) | |
|---|---|---|---|
| European Community | 267 | USA | 155 |
| Egypt | 100 | European Community | 104.5 |
| Morocco | 72 | Russia | 90 |
| Argentina | 70 | Brazil | 74 |
| Turkey | 60 | Saudi Arabia | 27 |
| Peru | 20 | Canada | 26 |
| Syria | 17 | Australia | 18 |

**Table 3. Main composition of some representative varieties of table olives from Spain (samples of 100 g of pulp have been used to calculate the composition) (Ministerio de Agricultura, Pesca y Alimentación, 2009).**

|  | Manzanilla | Hojiblanca | Gordal |
|---|---|---|---|
| Moisture (%) | 70 | 64 | 80 |
| Energy content (Kcal) | 210 | 207 | 102 |
| Fat content (%) | 21 | 20 | 9 |
| Proteins (g) | 1 | 1 | 1 |
| Fiber (g) | 1.5 | 2 | 1.5 |
| Minerals (g) | 4.5 | 4.5 | 5 |
| Carotens (mg) | 0.15 | 0.19 | 0.02 |
| C vitamin (mg) | 1.5 | 2 | 3 |
| Thiamine (mg) | 0.5 | 0.5 | 0.5 |

# 3. CONCLUSION

Table olives are one of the most appreciated fruits in the European Community, Spain being the major the producer and exporter of table olives in the world. Currently, although climatic conditions are different to those in the Mediterranean basin, table olives are being produced in many other regions throughout the world such as California, Australia, India, China, *etc*. Therefore, in the last two decades, the production of this fruit has almost doubled.

Table olives can be classified according to the physical aspect of the type of olive (green olives, olives turning color and black olives), trade preparation (treated olives, natural olives, dehydrated and/or shriveled olives, Olives darkened by oxidation and specialties), their style (whole olives, stoned (pitted) olives, stuffed olives, salad olives, olives with capers, olive paste, *etc*.), *etc*. In addition these fruits can be classified into three trade categories; extra (the best trade category), first class and second class (the poorest trade classification) depending on their percentage of defects. As table olives present a high concentration of *glycoside oleuropein*, it presents an unpleasant bitter taste, and because of this, olives cannot be eaten directly and need to be physicochemically treated. In most cases, the manufacturing process consists of three stages: firstly, table olives are harvested and transported to the factory. Here, lye treatment, washing and brining treatments are applied. Then, after the fermentation and storage period, the table olives are packed and in some cases pasteurized. Although in some cases the operational conditions of these stages are similar, these treatments should be defined depending on the type of olive. As all relevant aspects cannot be studied in detail here, scientific references are provided.

## REFERENCES

Agar, I. T., Hess-Pierce, B., Sourour, M. M. & Kader, A. A. (1998). Quality of Fruit and Oil of Black-Ripe Olives Is Influenced by Cultivar and Storage Period. *Journal of Agricultural and Food Chemistry, 46,* 3415-3421.

Arroyo-López, F. N., Durán-Quintana, M. C., Romero, C., Rodríguez-Gómez, F. & Garrido-Fernández A. (2007). Effect of storage process on the sugars, polyphenols, color and microbiological changes in cracked manzanilla-aloreña Table Olives. *Journal of Agricultural and Food Chemistry, 55,* 7434-7444.

Benítez, F. J., Acero, J. L., González, T. & García, J. (2001). Ozonation and Biodegradation Processes in Batch Reactors Treating Black Table Olives Washing Wastewaters. *Industrial & Engineering Chemistry Research, 40,* 3144-3151.

Boletín Oficial del Estado, Reglamentación Técnico Sanitaria para la elaboración, circulación y venta de las aceitunas de mesa. BOE 279, 2001, 42587- 42594.

Boskou, D. Olive oil: Chemistry and technology. 2nd Edition. Champaign, Illinois: AOCS PRESS; 2006.

Codex Alimentarius Commission. Codex Committee on processed fruits and vegetables. CX/PFV, 2004.

de Castro, A., Montaño, A., Casado, F. J., Sánchez, A. H. & Rejano, L. (2002). Utilization of Enterococcus casseliflavus and Lactobacillus pentosus as starter cultures for Spanish-style green olive fermentation. *Food Microbiology, 19,* 637-644.

International Olive Oil Council, world olive oil figures. (Last visited August 2009) URL http://www.internationaloliveoil.org/web/aa-ingles/corp/AreasActivitie/economics/AreasActivitie.html.

International Olive Oil Council. Quality management guide for the table olive industry. T.OT/Doc. no. 14. 2005.

International Olive Oil Council. Trade Standard applying to table olives. Res-2/91-IV/04. Madrid: IOOC. 2004.

Kailis, S. & Harris, D. Producing table olives. Stanley George Kallis. Collingwood Victoria, Australia. 2007.

López-López, A., Rodríguez-Gómez, F., Ruíz-Méndez, M. V., Cortés-Delgado, A. & Garrido-Fernández, A. (2009). Sterols, fatty alcohol and triterpenic alcohol changes during ripe table olive processing. *Food Chemistry, 117,* 127–134.

Lumaret, R., Ouazzani, N., Michaud, H., Vivier, G., Deguilloux, M. F. & Di Giusto, F. (2004). Allozyme variation of oleaster populations (wild olive tree) (Olea europaea L.) in the Mediterranean Basin. *Heredity, 92,* 343–351.

Maldonado, M. B., Zuritz, C. A. & Assof, M. V. (2008). Diffusion of glucose and sodium chloride in green olives during curing as affected by lye treatment. *Journal of Food Engineering, 84,* 224–230.

Medina, E., Brenes, M., Romero, C., García, A. & de Castro, A. (2007). Main antimicrobial compounds in table olives. *Journal of Agricultural and Food Chemistry, 55,* 9817–9823.

Ministerio de Agricultura, Pesca y Alimentación, Las aceitunas de mesa, Secretaria General de Agricultura y Alimentación, Dirección General de Industria Agroalimentaria y Alimentación, Spain (Last visited August 2009) URLhttp://www.castillalamancha.es/clmagro/pb/eventos2/archivos/24320081 449.pdf.

Montaño, A., Sánchez, A. H., Casado, F. J., de Castro, A. & Rejano, L. (2003). Chemical profile of industrially fermented green olives of different varieties. *Food Chemistry, 82,* 297–302.

Nanos, G. D., Kiritsakis, A. K. & Sfakiotakis, E. M. (2002). Preprocessing storage conditions for green 'Conservolea' and 'Chondrolia' table olives. *Postharvest Biology and Technology, 25,* 109-115.

Romero, C., Brenes, M., García, P., García, A. & Garrido, A. (2004). Polyphenol changes during fermentation of naturally Black olives. *Journal of Agricultural and Food Chemistry, 52,* 1973-1979.

Sánchez Gómez, A. H., García García, P. & Rejano Navarro, L. (2006). Elaboration of table olives. *Grasas y aceites, 57,* 86-94.

Sibbett, G. S., Ferguson, L., Coviello, J. L. & Lindstrand, M. Olive production manual. Oakland, California. University of California Agriculture and Natural Resources; 2005.

United Nations Conference on Trade and Development. International Agreement on Olive Oil and Table Olives, 2005 (Update: January 2009). Td/Olive Oil. 10/6.

Vega Macías, V. A., Rejano Navarro, L., Guzmán Díaz, J. P., Navarro García, C., Sánchez Gómez, A. H. & Díaz Montero, J. M. (2005). Recolección mecanizada de la aceituna de verdeo. *Agricultura. Revista Agropecuaria, 874,* 376-384.

# SECTION 3. OLIVE OIL

*Chapter 4*

# INDUSTRIAL METHODS TO PRODUCE EXTRA VIRGIN OLIVE OIL

## ABSTRACT

The manufacture of olive oil began *circa* five millenniums ago. From ancient methods to the most recent advanced technologies, the evolution of olive oil manufacture processes is described here. The first manufacturing equipment consisted of stone mortars, grinding olives and the atmosphere decantation process, but because of increased olive oil demand, new and more productive, sophisticated technology needed to be applied. This technology was notably improved with the arrival of electrical power, the hydraulic press, rolling wheels, centrifuges, *etc.* Manufacturing procedures, interesting practical advice, tips and scientific references related with the industrial production of extra virgin olive oil is provided here.

**Keywords**: History of olive oil manufacture; olive oil extraction; olive oil quality.

## 1. ANCIENT TECHNIQUES TO PRODUCE OLIVE OIL

The earliest references to the ancient techniques can be found in the Old Testament of the Bible (Micah 6:15).

*"You will put in seed, but you will not get in the grain; you will be crushing olives, but your bodies will not be rubbed with the oil; and you will get in the grapes, but you will have no wine." Micah 6:15*

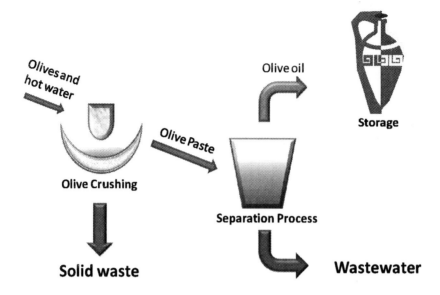

Figure 1. Flow diagram of olive oil manufacture during the Bronze Age (Kapellakis *et al.*, 2008).

As a consequence of the expansion of the olive tree, the olive oil extraction process was rapidly developed. During the Bronze Age (3,300 to 2,200 BC) in Crete, the first method used consisted of collecting the olives and squeezing them in stone mortars, the resulting paste was then collected in small pots and given that the water presented a higher density than the olive oil, the latter was separated using hot water, figure 1. This type of manufacturing process was mainly located in the home and operated by family members (Harwood and Aparicio, 2000; Kapellakis *et al.*, 2008). Ancient documents in Syria indicate that around 2,000 BC the value of olive oil was five times that of wine and two and a half times that of seed oils (Vossen, 2007). During the Mycenaean Era (*circa* 1,550 - 1,060 BC), the olive oil production method was slightly improved with respect to that used in the Bronze Age, figure 2. Linguistic indications have revealed that woven materials were used to extract the olive oil from the paste similar to those used by the Egyptians. This improvement consisted in introducing new equipment to complete the olive crushing (olive paste pressure process, figure 2) which consists of inputting the crushed olive (figure 2 shows after crushing) into woven material and, by pressure, the olive oil was separated from the olive paste. This process was used from the Mycenaean to the Hellenistic ages, and by using this improved method, a greater quantity of olive oil was extracted from the same amount of olive paste. However, at the end of this Era, due to the increasing trade of the

Roman Empire around the Mediterranean Sea, this improvement was still not efficient enough to meet the increased olive oil demand. Oil manufacturing passed from the family sphere to olive oil production plants, and then, other extraction processes with more production capacity were adopted. That is, it passed from a scale to satisfy family consumption (*vide infra*) to industrial production satisfying the necessities of the local population and international demand. In ancient times, olive oil was the most expensive edible oil, and although increasing productivity has meant a relative decrease in price, it still maintains this position.

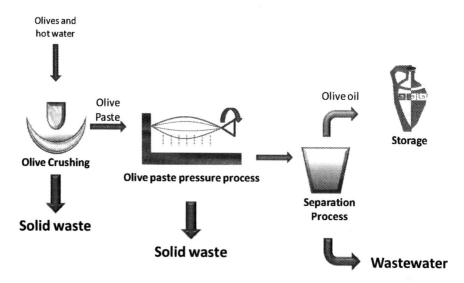

Figure 2. Flow diagram of olive oil manufacture during the Mycenaean Era (Kapellakis *et al.*, 2008).

In the Roman Empire, given the increased olive oil demand, the extraction process was further developed. Mainly, two of the first stages of the method used in the Mycenaean Era were optimized. As can be seen in figure 3, the olives were pressed by millstones, and then, the olive oil was separated by presses. In figure 4 olive presses from the Greek and Roman Empires are shown. As can be seen, the crushing of olives was carried out by the weight of the stone wheels and the wooden and iron tools used by the Greeks and Romans represented notable progress in olive oil production. Between the Roman and Venetian period, there were no outstanding improvements in olive oil production technology. The increased demand was met by increasing olive cultivation and production in occupied territories. From a technological point of view, the extraction process did not change for centuries. The next stage in the development was the invention of rolling wheels (moved by mules, donkeys or horses) and the invention of hydraulic pressing systems. During the twentieth century, the application of this equipment in olive oil extraction was carried out, figure 5 (Harwood and Aparicio, 2000; Kapellakis *et al.*, 2008). However, as demand continued to increase, production of olive oil was not sufficient, and other techniques based on malaxation of mill olives and centrifugations of olive paste appeared and were applied industrially. These new technologies have been classified and described here as modern techniques to produce olive oil.

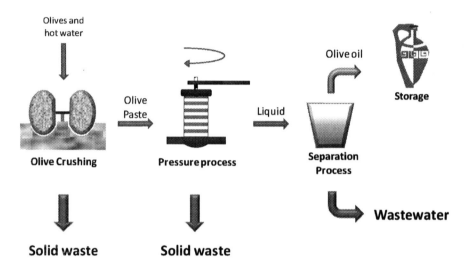

Figure 3. Flow diagram of olive oil manufacture during the Roman Era (Kapellakis *et al.*, 2008).

Figure 4. Ancient olive mills. (a) Used in ancient Greek olive oil production (Kilizman, Turkey); (b) used in Pompeii, 79 AC.

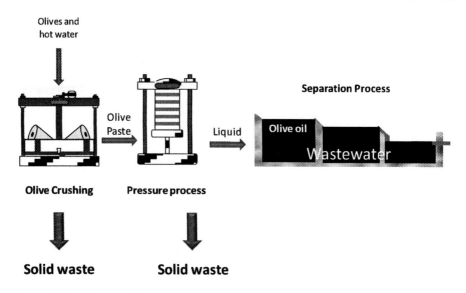

Figure 5. Flow diagram of olive oil manufacture in the twentieth century (Kapellakis *et al.*, 2008).

## 2. MODERN TECHNIQUES TO PRODUCE OLIVE OIL

During the 2008/09 season, more than $2.8 \cdot 10^6$ tons of olive oil will be produced in the world and 75 % of this amount will be produced in the European Community (International Olive Council, 2009). The annual production of olive oil during the last 18 years in the world and the European Commission figures are shown in figure 6. The curve passes through consecutive maximums and minimums, but in between highs (During 1991/92, 1996/97, 2003/04 seasons) there were years of lower production (During 1990/91, 1995/96, 2002/03 seasons). One of the factors causing this fluctuation is that the trees require periods of recuperation after years of particularly high yields (see Chapter 2). Even though modern cultivation techniques are reducing these periods of lower yields, it is probable that these cycles will continue to exist. The harvesting periods depend on regional and weather conditions, and in all cases, olives are alternate year bearing fruits (Sibbet *et al.*, 2006). During the last two decades, following a general trend, olive oil production has increased. Given that olive oil is being produced in more and more non European countries, relatively production in non European countries increases by 11 % compared with European countries production, as can be seen in figure 6.

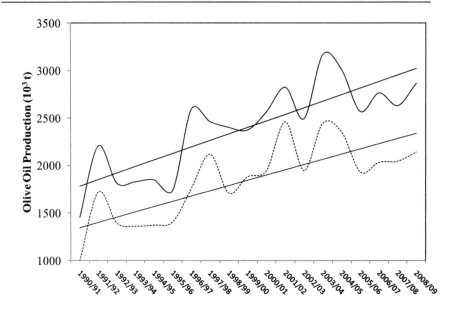

Figure 6. Olive oil production in the World (solid line) and the European Union (dotted line). Both linear productivity tendencies are represented here by the thinner lines (International Olive Council, 2008).

The olive and olive oil industry is an important employer in the agro-food-sector with employing over 800,000 people in Europe alone, encompassing approximately 600 companies in the European Union (Technology Dissemination Centres, 2009). During the 2006/07 and 2007/08, depending on the harvest region, the price of extra virgin olive oils was approximately constant ranging between 2.3 and 3.1 €·kg⁻¹. Nevertheless, during the 2008/09 season, the price of extra virgin olive oil has decreased to 1.7 and 2.1 €·kg⁻¹ (International Olive Council, 2009). This fall in prices represents a reduction of more than 30 % in income, and therefore will affect the development of the olive oil sector. Although this fall in price is present in all sub sectors related with the olive oil sector (refined olive oil and refined olive pomace oil), the extra virgin olive oil trade is the most sensitive and therefore the most affected.

The main objective of the olive oil manufacturing process is to extract the greatest possible amount of high quality olive oil without organoleptic alteration. In general, the olive oil production process requires the separation of the oil phase from the mixture formed by skin, pulp, kernel, water and oil phase. Due to the fact that olives are fruits and its oil is made only by mechanical or physical means, it can be considered as an oily juice of fruits. This differs fundamentally from other edible oils which require chemical reactions for their production.

One of the biggest handicaps is that more than three million of tons of olives must be dealt with in less than five months. Once the olives have been collected at the point of optimum maturity, olive oil extraction must be done as soon as possible to obtain olive oil of the highest quality (Torrecilla *et al.*, 2006). The four most modern techniques to manufacture extra virgin olive oil are the pressure method (discontinuous process), the three-phase and two-phase techniques and percolation/centrifugation (continuous processes). The two phase and three phase methods are shown in figure 7. The main difference between these consists of the centrifugal equipment used to separate the olive oil from its paste (*vide infra*). In order to obtain olive oil of the highest quality, the olives should be correctly harvested, transported, *pre*-treated, crushed, mixed, and then, the olive oil is separated from the olive cake by centrifugation. Finally, vertical centrifuging or sedimentation is used to purify the olive oil, figure 7. (Atti, 1926; Boskou, 2006; Kapellakis *et al.*, 2008; Torrecilla, 2001; Torrecilla, 2004).

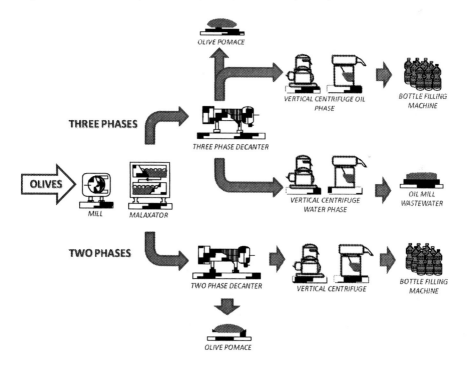

Figure 7. Flow diagram of current olive oil manufacture called two and three phases.

## Harvesting

The most important aspect at this stage consists of collecting the fruits in an ideal state of maturity. The green color of the olive is the usual indicator of its ripeness. When the color has changed to black this means that no more oil will be produced, and harvesting should take place as soon as possible after this because of the possibility of autumn/winter rains which would increase the water content in the fruit. In general, in countries of the Mediterranean region and from California, the harvest period is between November and January and September and November, respectively. There are two groups of harvesting methods, viz. by hand and mechanical means (Sibbet et al., 2005; Grattan et al., 2006).

- **Hand harvesting** is the most expensive method due to high labor costs. To acquire ideal access to the olives, the pruning of the canopy requires specific considerations. The fruits are removed by sliding the cupped, gloved hand down the shoot in a milking action (Sibbet et al., 2005; Gucci and Cantini, 2008).
- **Mechanical Harvesting** falls into two general categories, viz. shaking the trunk or branches or by picking heads, which connect directly with the canopy (Ferguson, 2006). In both cases, the removal efficiency is 90 % which can be increased when harvesting mature olives. Usually, this type of harvesting produces more damage than hand harvesting (Ferguson, 2006; Sibbet et al., 2005).

In both cases, to maintain the highest olive oil quality, the harvesting method must not damage the fruit and this cannot be contaminated by soil material. In general, the pruning processes of young and mature olive trees are usually focused on harvesting by mechanical methods. In fact, new orchards are designed to be harvested mechanically. Once the olives are harvested, and waiting for their juice to be extracted, they should be spread out and to allow good ventilation, storage in sacks is not adequate (Boskou, 2006).

## Transport

Care should be taken during transport from orchards to factory regarding the environmental temperature, mechanical damage and handling of the olive to minimize possible defects in the fruits. The best method to transport olives is in open mesh plastic crates that allow air circulation, which means that damage from excessive heat is minimum, thus inhibiting their catabolic activity. At this stage, improper handling of the olives could initiate the degradation by incipient enzymatic reactions, affecting the organoleptic characteristic and the quality of the olive oil produced (Boskou, 2006).

## Pre-Treatments

Once the olives are in the oleic factory, these are transported to *pre*-treatment equipment on a moving belt, where leaves and light materials are removed by a powerful air flow. The olives are then washed by wet cleaning equipment to remove all foreign material which could be harmful to the equipment or contaminate the final olive oil. Whenever possible, once the non desirable materials have been removed, this water should be recycled. This stage is important because if this process is not adequately carried out, the final olive oil could present an unpleasant bitter taste.

## Crushing

The main objective of this stage is to transform the collected olives into olive paste, and when they are adequately clean, these are crushed. Usually, this operation is carried out by stone mills or hammer mills which work in a continuous flow system. The stone mill is cylindrical with a diameter between 1.2 to 1.4 m with a side shutter to allow the discharge of the olive paste. It is made up of 2 or 4 millstones which gyrate at 12 - 15 rpm. The performance of this equipment is better than other types, because it combines the pressing and pushing actions, and then, pre-mixes the produced paste (Boskou, 2006). At present, in some traditional stone mills, the continuous extraction process is being applied. In the case of hammer mills, the hammers drive the olives against breaker plates (a fixed or slowly gyrating metal screen) and the size reduction of the olives is due to impact and attrition forces. The screen has holes of 5, 6 or 7 mm depending on the posterior extraction process and the ripeness of the olives. In order to maintain constant operating conditions, a variable quantity of water is introduced into the equipment. The hammers of the mill can be replaced by cutters, bars, *etc.* (Brennan *et al.*, 1990). Other types of mills can also be used (stone mills, disc mills, *etc.*). Currently there many disagreements regarding the most appropriate mill (Boskou, 2006). Although the hammer mill requires higher malaxation times, and due to rubbing, the temperature of the olive paste is raised, resulting in an increase in the bitterness of the olive oil, and it also gives a greater handling capacity, is faster, occupies less space, facilitates cleaning and is less expensive.

## Malaxation

The goal of this stage is to gather together all the individual drops of oil (20-25 % of the olive) breaking up the oil/water emulsion. The malaxation (also called beating or kneading) stage is used to reverse the homogenization process and frees the olive oil to allow for easier extraction later and to increase the amount of available olive oil. The relationship between the dimension of the initial droplet and subsequent oil phase can be seen in Table 1.

**Table 1. Percentages of oil drop diameters after of crushing and malaxation processes (Boskou, 2006).**

|                  | <15 μm | 15-30 μm | 30-45 μm | 45-75 μm | 75-150 μm | >150 μm |
|------------------|--------|----------|----------|----------|-----------|---------|
| After crushing   | 6      | 49       | 21       | 14       | 4         | 6       |
| After malaxation | 2      | 18       | 18       | 18       | 19        | 25      |
| Increasing rate  | -4     | -31      | -3       | 4        | 15        | 19      |

Not only does the action of mixing have a direct effect on the nutritional quality, but it also has a significant effect on the sensory qualities (Fellow, 1998). In the malaxation and previous stages, aromas are created by the action of fruit enzymes. Malaxators consist of cylindrical vats with rotating blades which revolve slowly, spinning at 15-20 rpm (less than 30 rpm) and double walls to maintain the temperature of the olive paste at less than 30 °C. Depending on the maturity of the olive, the paste is mixed for about 20-30 minutes (never more than 90 minutes). When millstones have been used in the crushing stage, the malaxation process takes 15-20 minutes at room temperature. These conditions have been established to preserve the volatile aromatic compounds. In general, this equipment is made of stainless steel to prevent the oxidation of the cake and nitrogen, as an inert gas, is sometimes used to aid this prevention. This equipment can be employed in a vertical or horizontal position adapting to the space available. With some varieties, and also depending on the maturity of the olives it is necessary to apply longer period of time and higher temperatures of the malaxation, in order to obtain higher olive oil extraction yields. (Boskou, 2006; Kapellakis et al., 2008).

## Olive Oil Extraction Processes

The main objective of this stage is to split the oily juice of the olive from the olive paste. The mixed cake is composed of skin, pulp and pits. The principal constituents of the pulp are water and oil whose percentages are somewhat less than 60 % and 30 % respectively (Torrecilla, 2004; Aragón and Palancar, 2001). Once the paste has been adequately mixed, four methods can be applied to extract its oil content *viz.* pressure (discontinuous process), two centrifugation processes, or percolation/centrifugation processes.

**Pressure process.** Comparing figures 4 and 5, the main difference between the olive oil manufacturing processes carried out in the era of the Roman Empire and the twentieth century involves the technology used and the power source, i.e., in the Roman Era, as opposed to today, the energy necessary to crush olives and to carry out the pressure process on the olive paste was animal driven. This method is still widely used today, and it is still a valid way of producing high quality olive oil.

The pressure process consists of spreading the olive cake on oil diaphragms (less than 3 cm thick), which are stacked one on top of the other. Traditionally, these disks were made of hemp fibers, but nowadays they are made of synthetic fibers which are easier to clean and maintain. They are placed in moving units with a central shaft. In order to obtain a homogeneous pressure on the cake, metal trays are inserted between every three or four oil diagrams. These disks are then put onto a hydraulic piston, forming a pile and pressure is applied to the disks (less than 400 bars), thus compacting the solid phase of the olive paste and percolating the oil and water phases running through the olive cake (Kapellakis *et al.*, 2008). The applied pressure increases gradually, reaching a final pressure after 45-60 min and is maintained during 10-20 min. For moistures of less than 30 %, the processing yield is about 85-90 %. After each extraction, the disks must be properly cleaned, because fermentation processes could be triggered by the leftover paste, and then, the olive oil produced subsequently would be contaminated. In this classical method, once the olive oil and water are separated from the olive cake, to separate the olive oil from the olive oil and water mixture, a vertical centrifuge (*vide infra*) and/or standard decanting processes (separation process shown in figure 5) are used.

**Centrifugation processes,** most of the olive oil in the world is being produced using this type of process. In relation to the pressure process, the main difference is the operating plant mode; the centrifugal processes are continuous, and therefore, the processing capacity is higher and costs are less. Given that the olive oil, water and solids present different density values, the separation of these compounds from the mixer paste can be carried out by centrifugal force. In this type of process, forces 3,000 times greater than normal gravitational force are applied. The equipment used is a horizontal centrifuge, also called a decanter. This consists of a cylindrical conical bowl that rotates at about 3,000 rpm, and inside there is a coil which rotates at a slightly lower speed, pushing the material out of the decanter (Brennan *et al.*, 1990). The action of the decanter causes the pomace to move to one end of the centrifuge, while the olive oil and olive mill wastewater (OMW) are pushed to the other end. At this point, the OMW and olive oil can be separated. Increasing the speed, higher productivity per unit of time can be achieved. Currently, there are two decanter types which carry out the centrifugal processes, *viz.* three phase and two phase decanters. With slight modifications the same decanter can operate in both.

- *Three-phase centrifugation.* This type of decanter was the first equipment used to separate olive oil from other compounds. And is so called because of its three outputs, these being olive oil, water (added and vegetable water) and wet solid. In order to achieve a satisfactory performance with this equipment, water, equivalent to 50 % of the total weight of olives, needs to be added. In general, given that around 1.5 times more water than in press extraction is necessary, a larger quantity of wastewater requires treatment, and also, the olive oil has less valuable natural antioxidants which dissolve and are lost in the aqueous phase. An effective solution to these problems consists of recycling the water used, and in this way, its use is reduced by 35 % and therefore polyphenols can be recovered. However, this improvement negatively affects the quality of the olive oil produced and as a consequence, a new decanter type based on two phase centrifugation was proposed.
- *Two-phase centrifugation.* This type of decanter is attractive particularly in Spain, because additional water is unnecessary. However, when the raw material consists of mainly dry olives, the addition of small quantities of water could be necessary. This equipment has two outputs viz. one for olive oil and the other for wet olive pomace (*alpeorujo* in Spanish). The latter is composed of water (mainly vegetable water, moisture higher than 63 %) and solids (Torrecilla, 2004; Torrecilla 2001). Olive pomace is a non-Newtonian fluid with characteristics of either viscoelastic or Binghan-plastic materials depending on the water content and the strain conditions inside a given apparatus. And therefore, this solid-liquid waste is complicated to handle and more difficult to treat industrially (Torrecilla *et al.*, 2006). This olive oil is green and has a

higher concentration of aliphatic alcohols. In general, the productivity yields of both continuous techniques are similar, but, given the higher concentration of natural antioxidants, the quality of olive oil produced using this decanter is superior than that produced by the former.

**Table 2. Main characteristics of the four types of olive oil extraction processes.**

|  | Pressuring process | Three phase | Two phase | Percolation |
|---|---|---|---|---|
| **Process** | Discontinuous | Continuous | | |
| **Capacity** | Small | Medium High | | |
| **Labor cost** | High | Low | | |
| **Energy consumption** | Low | High | | |
| **Water consumption** | Low | High | Low | High |
| **Processing yields (%)** | 86-90 | 85-89 | | |
| **Pomace moisture (%wb)** | 28 | 48 | 60 | 48 |
| **Olive mill waste water quantity** | Low | High | Low | High |
| **Polyphenol content** | High | Low | High | |
| **Contamination Risk** | High | Low | | |

**Table 3. Advantages and disadvantages of the manufacturing process of extra virgin olive oil.**

| Systems | Advantages | Disadvantages |
|---|---|---|
| Pressure | Less energy required. The produced oil relatively drier and easier to manage. Reduction of the added water. | Less capacity. More labor intensive. Difficult to maintain cleanliness of oil diaphragms. More paste and oil contact with oxygen. Non continuous process. Longer period from harvest to pressing. |
| Three-phase decanter | Continuous and automated process. Produces relatively dry pomace. Compact machinery. Limited labor required. Better yield performance. | Expensive. More technical labor is required High energy consumption. Reduced antioxidant concentration. Water content. Increase in the production of OMW. |
| Two-phase decanter | Continuous and automated process. Compact machinery. Limited labor required. Less water required. Oil produced contains more phenols with more aroma and more resistant to oxidation. Uses only one vertical centrifuge. Better yield performance. | Expensive. More technical labor is required. High energy consumption. Produces a very wet pomace. |
| Percolation/centrifugation process | Higher polyphenol content of olive oil. Environmental temperature method. Low labor. Low energy requirement. | Need for more space. Large surface areas can lead to rapid oxidation of the olive oil. |

Taking into account the main characteristics of the different methods to produce olive oil *viz.* pressure, three and two phase decanters, their main advantages and disadvantages are shown in tables 2 and 3. Their influence on the quality of the olive oil produced will be treated in Chapter 5. As a quantitative reference, the quantity of waste and olive oil produced using 1,000 Kg of olives is shown in table 4. In both systems, all liquid phases are sent to a vertical centrifuge which operates at 6,000 rpm and is used to separate small quantities of water from the olive oil. In some cases, warm water is added to clean the fine solids contained in the olive oil.

**Table 4. Comparative data for olive oil extraction processes, for 1,000 kg of olives. (Aragón and Palancar, 2001).**

| Process | Input | Input amount | Output | Output amount |
|---|---|---|---|---|
| Pressing | Olives | 1000 kg | Oil | 200 kg |
| | Washing Water | 0.1-0.12 m$^3$ | Solid Waste (25 % water + 6 % oil) | 400 kg |
| | Energy | 40-63 kWh | Waste water (88 % water) | 600 kg |
| Three-phase | Olives | 1000 kg | Oil | 200 kg |
| | Washing Water | 0.1-0.12 m$^3$ | Solid Waste (50 % water + 4 % oil) | 500-600 kg |
| | Fresh water for decanter | 0.5-1 m$^3$ | Waste water (94 % water + 1 % oil) | 1,000-1,200 kg |
| | Water to clean the impure oil | 10 kg | | |
| | Energy | 90-117 kWh | | |
| Two-phase | Olives | 1,000 kg | Oil · | 200 kg |
| | Washing Water | 0.1-0.12 m$^3$ | Solid Waste (60 % water + 3 % oil) | 800-950 kg |
| | Energy | < 90-117 kWh | | |

**Percolation/centrifugation processes**. To produce olive oil of high quality, the percolation process and centrifugal process can be combined (Sinolea system). Different percolation extractors called "Alfin", "Acapulco" and "Acapulco-Quintanilla" were used prior to 1972, when the Sinolea extractor was introduced. This is based on semicylindrical grating and hundreds of small stainless steel discs or plates which are dipped into the paste; the olive oil sticks to the metal and is removed from the paste by scrapers in a continuous process. This first stage takes from 30 to 60 min. Given that this process is not 100 % efficient because it leaves a large quantity of oil in the olive paste, the remaining paste has to be reprocessed by centrifugal processes. In some cases, prior to centrifugal process, the malaxation process is required. Although the two pieces of equipment are combined, the percolation/centrifugal processes are continuous, and then, the processing capacity is greater than when using a decanter. In addition, due to the content of polyphenols of the olive oil, this presents similar quality characteristics to that which is produced by pressure methods (Boskou, 2006).

## Clarification

In the aforementioned four systems, all liquid phases produced are sent to a clarifier or settling tank.

**Sedimentation.** This process is used to improve the olive oil quality. In the best cases, in the aforementioned four processes (pressure, three-phase, and two-phase methods and percolation/centrifugation processes), the cleaned olive oil still contains less than 0.1 % of water or solids, which should be removed. As the olive oil, water and solids present different densities, these compounds can be easily separated and it is carried out by sedimentation tanks at temperatures of less than 16 °C, where the olive oil settles for about 2 months. Currently, for modern industrial processes, these sedimentation times are unsuitable

**Clarifier.** This equipment is a vertical disc stack centrifuge which is used to separate small quantities of water from the olive oil and can work continuously. Generally, it operates at about 6,000 rpm which produces a centrifugal force higher than 10,000 times the gravitational force. And therefore, the settling time is lower than in the case of the sedimentation process.

## 3. CONCLUSION

From ancient methods to recent technologies, the evolution of olive oil manufacturing processes has been here described. As a consequence of the number of people demanding products related with the olive, olive oil manufacture has passed from family to industrial spheres. And as during ancient ages, the demand for the olive oil is still increasing, and the designers of their manufacturing processes are continuously trying to find solutions to help meet these requirements. The manufacturing process has passed from stone morters, grinding olives and the atmosphere decantation process to hammer mills, malaxation process and centrifugal decanters. The historical olive trade evolution is a clear example of the improvement of manufacturing technology to satisfy the specific world market necessities in the sense of productivity and the characteristics of quality.

## REFERENCES

Aragón, J. M., & Palancar, M. C. IMPROLIVE 2000, present and future of alpeorujo. Madrid: Ed. Complutense S. A.; 2001.
Atti, M. D. (1926). New method of olive oil extraction. *Journal of the American Oil Chemists' Society, 3*, 401-402.
Boskou, D. Olive oil: Chemistry and technology. 2nd Edition. Champaign, Illinois: AOCS PRESS; 2006.
Brennan, J. G., Butters, J. R., Cowell, N. D. & Lilley, A. E. V. Food Engineering Operations. 3rd edition. Great Yarmouth: Elsevier Science; 1990.
Fellow, P. Food Processing technology principles and practice. Cambridge: Woodhead publishing limited; 1998.
Ferguson, L. (2006). Trends in Olive Harvesting. *Grasas y Aceites, 57*, 1-7.
Grattan, S. R., Berenguer, M. J., Connell, J. H., Polito, V. S. & Vossen, P. M. (2006). Olive oil production as influenced by different quantities of applied water. *Agricultural Water Management, 85*, 133-140.
Gucci, R. & Cantini, C. Prunning and training systems for modern olive growing. Collingwood, Australia. CSIRO publishing; 2008.
Harwood, J. & Aparicio, R. Handbook of olive oil: Analysis and properties. Gaithersburg, Maryland. Aspen Publishers, Inc. 2000.
International Olive Oil Council. 2009. (Last visited August 2009) URL http://www.internationaloliveoil.org.
Kapellakis, I. E., Tsagarakis, K. P. & Crowther, J. C. (2008). Olive oil history, production and by-product management. *Reviews in Environmental Science and Biotechnology, 7*, 1–26.
Sibbett, G. S., Ferguson, L., Coviello, J. L. & Lindstrand, M. Olive production manual. Oakland, California. University of California Agriculture and Natural Resources; 2005.
Technology Dissemination Centres. By-Product Resusing from olive and olive oil production. 2004. (Last visited August 2009) URL http://www.biomatnet.org/publications/1859bp.pdf.
Torrecilla, J. S. (2001). Aprovechamiento del alpeorujo. *Agricultura, revista agropecuaria, 832*, 734-737.
Torrecilla, J. S. Secado del orujo en lecho fluidizado movil. Madrid: Ed. Complutense S.A.; 2004.

Torrecilla, J. S., Aragón, J. M. & Palancar, M. C. (2006). Improvement of fluidized bed dryers for drying olive oil mill solid waste (olive pomace). *European Journal of Lipid Science and Technology, 108*, 913–924.

Vossen, P. (2007). Olive Oil: History, Production, and Characteristics of the World's Classic Oils. *Hortscience, 42*, 1093-1100.

*Chapter 5*

# VIRGIN OLIVE OIL QUALITY

## ABSTRACT

Regulation, parameters and chemical concentration related with the quantification of olive oil quality are described here. The two most important classifications of olive oils in the world are also summarized. The first provided in 1948 by the United States Department of Agriculture, and the second given by the prestigious International Olive Oil Council (IOOC), Codex Alimentarius and the European Commission. As nearly all members of IOOC are from the Mediterranean basin, and as this classification is the most recent, the latter classification is currently the most used in Europe. Focusing on extra virgin olive oil manufacture, the aspects with the most influence on the quality of the oily juice of the olive is also presented. In addition, relevant factors of interest to consumers regarding the quality of the olive oil have also been studied. Interesting tips and scientific references are provided here.

**Keywords**: Adulteration; quality of olive oil; Influence of olive oil manufacture on its quality; Categories of olive oil.

## 1. INTRODUCTION

There are three methods to assess the quality of foodstuffs; one depends on the human senses, principally taste, one uses physicochemical parameters and the third is a combination of both. Taking the first into account, for any given food, many qualitative evaluations can be found. That is, the organoleptic evaluation can be made by the evaluation of experts tasters (checking the flavor of something by eating) and by physicochemical variables.

The adulteration of food products with cheaper and more available substitutes is a worldwide problem which has existed for centuries. Currently, adulteration of foods is more and more prevalent, mainly in those products with relatively high prices. Given that the highest quality olive oil (extra virgin olive oil, *vide infra*) is in this category, a large number of cases of adulteration of this oily juice have been detected in recent years. The substitution or adulteration of extra virgin olive oil with cheaper ingredients is not only an economic fraud, but may also on occasion have severe health implications for consumers. An example being the Spanish toxic oil syndrome resulting from the consumption of aniline denaturalized rapeseed oil that involved more than 20,000 people causing serious illness and even death (Lee *et al.*, 1998; Mildner-Szkudlarz and Jelen, 2008). To fight against the increase of these fraudulent activities, the chemical compositions of specific olive oils have been qualified and protected by denominations of origin. Technically, mathematical algorithms (Fonseca *et al.*, 2006; Marini *et al.*, 2007; Torrecilla *et al.*, 2009), physicochemical parameters, indexes, *etc.* have been also proposed (Lerma *et al.*, 2008; Sayago *et al.*, 2007; Angiuli *et al.*, 2006).

Extra virgin olive oil is subject mainly to two types of adulteration. In the first group, the extra virgin olive oil is blended with low-grade olive oils (olive-pomace oil, virgin olive oil obtained by a second centrifugation of the olives or refined olive oils). The second consists of mixing extra virgin olive oil with other oils such as seed oils (hazelnut, sunflower, maize, corn, soybean, palm, *etc.*) which, although similar, are substantially cheaper (Mildner-Szkudlarz and Jelen, 2008; Peña *et al.*, 2005; Dourtoglou *et al.*, 2003). Given the chemical similarities of extra virgin olive oil and hazelnut oil, this adulteration is difficult to detect, especially when its concentration is low (Torrecilla *et al.*, 2009; Mildner-Szkudlarz and Jelen, 2008; Peña *et al.*, 2005).

## 2. OLIVE OIL QUALITY

In general terms, the quality of foods is defined by Gould as "*The combination of attributes or characteristics of a product that have significance in determining the degree of acceptability of that product by the user*" (Gould, 1992). Applying this general description to olive oil, its quality can be divided into two parts *viz.* the nutritional perspective and the organoleptic (Duran, 1990). High nutritional values and pleasant taste contribute equally to the increasing demand for olive oil. From the point of view of nutrition, high values of oleic acid and a low content in phenolic compounds are necessary. Regarding the organoleptic factor, the presence or absence of volatile compounds is a good indicator of olive oil quality (Kalua, 2007; Vossen, 2007). In general, to quantify the quality of olive oil, the International Olive Oil Council bases this on chemical parameters such as the free fatty acid content, peroxide value, ultraviolet specific extinction coefficient at 232 and 273 nm or sensory score (International Olive Council, 2008). Due to hydrolytic and oxidative reactions (Duran, 1990), the parameters of olive oil quality change between the time the olive oil is produced and its consumption by the user (*vide infra*). The degradation of olive oil compounds by the lipase enzyme is schematically shown in figure 1.

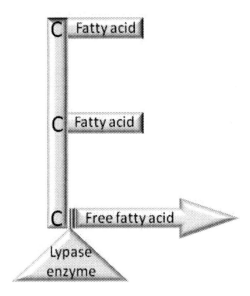

Figure 1. Hydrolysis process of a triglyceride by the action of the lipase enzyme

## 2.1. Categories of Olive Oil

To establish the quality of all olive oil categories, quantitative characteristics of quality have been defined. Firstly, the possible types of olive oil should be defined. There are two main classifications of olive oils in the world. One has been defined by the United States Department of Agriculture (USDA) and the other by The International Olive Oil Council (IOOC).

- **United States Department of Agriculture** currently lists four grades of olive oil. These grades were published in the Federal Register of February 20, 1948, to become effective March 22, 1948. These are based on acidity, absence of defects and sensory score (odor and flavor), the main characteristics are shown in table 1 (United States Department of Agriculture, 1948). In March 28, 2008, an additional and more extensive classification was published as a draft (United States Department of Agriculture, 2008).
- **The International Olive Oil Council** is the world's only international intergovernmental organization in the field of olive oil and table olives. It was set up in Madrid, Spain, in 1959, under the auspices of the United Nations. The Council is a decisive player in contributing to the sustainable and responsible development of olive growing and it serves as a world forum for discussing policymaking issues and tackling present and future challenges. This organization along with the European Commission (EC) and Codex Alimentarius has established nine different categories of olive oil. These categories with their main chemical characteristics are shown in Table 2. The European regulations do not permit the trade of either the refined olive oil or refined olive pomace oil. But trading in their blending with virgin olive oils is permitted.

**Table 1. Categories of olive oils according to United States Department of Agriculture. (United States Department of Agriculture, 1948)**

| Grades of olive oil | Characteristics |
|---|---|
| U.S. Grade A or U.S. Fancy | The olive oil is free from defects, its typical color is between greenish to pale yellow. Free fatty acid ≤ 1.4 % Sensory score ≥ 90 |
| U.S. Grade B or U.S. Choice | The olive oil is reasonably free from defects, its typical color is between greenish to pale yellow. Free fatty acid ≤ 2.5 % Sensory score ≥ 80 |
| U.S. Grade C or U.S. Standard | The olive oil is fairly free from defects, its typical color is between greenish to pale yellow. Free fatty acid ≤ 3 % Sensory score ≥ 70 |
| U.S. Grade D or Substandard | is the quality of olive oil that fails to meet the requirements of U.S. Grade C or U.S. Standard. |

**Table 2. Categories of olive oils according to The International Olive Oil Council, European Commission and Codex Alimentarius.**

| Category | | | Characteristics |
|---|---|---|---|
| Virgin olive oil | Edible virgin olive oil | Extra virgin olive oil | Oleic acid $\leq 0.8$ % <br> Sensory score $\geq 6.5$ <br> $K_{232} \leq 2.50$ <br> $K_{270} \leq 0.22$ <br> $\Delta K \leq 0.01$ |
| | | Virgin olive oil | Oleic acid $\leq 2.0$ % <br> Sensory score $\geq 5.5$ <br> $K_{232} \leq 2.60$ <br> $K_{270} \leq 0.25$ <br> $\Delta K \leq 0.01$ |
| | | Ordinary virgin olive oil | Oleic acid $\leq 3.3$ % <br> Sensory score $\geq 3.5$ <br> $K_{270} \leq 0.30$ <br> $\Delta K \leq 0.01$ |
| | Non edible virgin olive oil | Lampante virgin olive oil | Oleic acid $> 3.3$ % <br> Sensory score $\leq 3.5$ |
| Refined olive oil * | | | Oleic acid $\leq 0.3$ % <br> $K_{270} \leq 1.10$ <br> $\Delta K \leq 0.16$ |
| Olive oil ** | | | Oleic acid $\leq 1.0$ % <br> $K_{270} \leq 0.90$ <br> $\Delta K \leq 0.15$ |
| Olive pomace oil | Crude Olive pomace oil # | | No limit |
| | Refined Olive pomace oil ## | | Oleic acid $\leq 0.3$ % <br> $K_{270} \leq 2.00$ <br> $\Delta K \leq 0.20$ |
| | Olive pomace oil ### | | Oleic acid $\leq 1.0$ % <br> $K_{270} \leq 1.70$ <br> $\Delta K \leq 0.18$ |

\*     Olive oil mainly obtained from the lampante virgin olive oil.
\*\*    Olive oil obtained by the blending of refined olive oil and virgin olive oil (edible oil).
\#     Olive oil obtained from pomace olive (edible oil)
\#\#   Olive oil obtained from crude olive oil and refining process
\#\#\# Olive oil obtained by blending refined olive pomace oil and virgin olive oil (edible oil)

## 2.2. Chemical Composition of the Olive Oil

The sensory features of olive oil vary depending on its chemical composition, as an example, some organoleptic defects and the volatile compounds responsible are shown in table 3. In the light of recent EC legislation and researches developed, there are some parameters that quantify the quality of olive oil such as free acidity, chemical composition, UV absorbance, peroxide value, *etc*. The free acidity (free fatty acids) is the oldest parameter in the evaluation of olive oil quality. The free acidity of olive oil, in the form of oleic acid, gives us an indicator to its quality and stability. This parameter is used to classify a given olive oil within a specific category of olive oil, Tables 1 and 2. Nevertheless, as a higher concentration of free acids has greater relevance depending on the presence of other constituents such as polar phenols, whose concentrations vary with the hydrolytic and oxidative processes. Therefore, in order to provide reliable quality parameters, the free acidity and the chemical composition should be analyzed as a whole (Boskou, 2006). Other important parameters to evaluate the quality of olive oil are the isomers of diacylglycerols concentration. In particular, the ratio of 1, 3-diacylglycerols versus 1, 2–diacylglycerols or this latter compound against total diacylglycerols isomers (Torrecilla *et al.*, 2009; Boskou, 2006) are adequate ratios to describe the quality of olive oil taking into account its storage features.

**Table 3. Chemical compounds responsible for sensory defects of olive oil (Boskou, 2006)**

| Sensory defects | Causes |
|---|---|
| Mustiness, humidity | 1-octen-3-ol. |
| Fusty sensory defect | Ethyl butanoate; propanoic and butanoic. |
| Winey vineagary | Acetic acid; 3-methyl butanol and ethyl acetate. |
| Rancid defect | Saturated and unsaturated aldehydes and acids. |
| High level of oxidative alterations | Acetic; butanoic; hexanoic and heptanoic acids. |
| Rancid flavor | 2-octenal; 2-heptanal and 2-decenal. |
| Final aroma | Hexanal; nonanal; octanal; pentanal and heptanal. |

UV absorbance values at 232 and 270 nm are respectively described by $K_{232}$ and $K_{270}$ and $\Delta K$ value is defined by equation 1 below. These parameters are useful to determine the first stage of oxidation of extra virgin olive oil and the complete oxidation process during its storage periods (Bilancia *et al.*, 2007). In addition, the peroxide value is an important parameter which can be used to qualify the oily juice. The latter and $K_{232}$ values can evaluate the influence of the changes on the lipid content of olive oil, and then on its quality (Vekiari *et al.*, 2007).

$$\Delta_K = K_{\max} - \frac{K_{\max+4} + K_{\max-4}}{2}$$

(1)

In equation 1, $K_{max}$, $K_{max+4}$ and $K_{max-4}$ are the maximum absorbance near 270 nm and the absorbance at four nm approximately, respectively. In some cases, some aforementioned parameters could be correlated, that is, these related parameters give the same information related with the oxidation of olive oil. For example, when the oxidative status and antioxidative capacity are being considered, the peroxide value is correlated with the $K_{232}$ and total phenol content. Here, as $K_{232}$ was selected because it gives us more information. In general, in these cases, the most informative parameter is selected (Grigoriadou and Tsimidou, 2006).

During the production process of olive oil (mainly in the olive ripening, blending, extraction and storage processes, *vide infra*) enzymes such as lipase, lipoxygenase, phenoloxidase, hydroperoxide, lyase/isomerase, *etc.* are produced. The activities of these enzymes negatively affect olive oil quality, figure 1. However, the activity of lipoxygenase and other enzymes increase the sensory features of olive oil. During extra virgin olive oil manufacture, some of these enzymes pass to the olive pomace, and then to the olive pomace oil, facilitating its deterioration throughout its storage process. On the other hand, it is not clear if the microbiological activity improves, or not, the sensory features of olive oil. Ciafardine and Zullo suggested two apparently opposing statements. Although the microbiological and enzymatic activity decreases the bitter characteristics of olive oil, the highest olive oil quality is achieved when the microorganisms are completely removed (Ciafardine and Zullo, 2002; Ciafardine and Zullo, 2002a).

In addition to free acidity, peroxide value and the parameters related with the UV absorbance values, there are other organoleptic defects of olive oil which are not quantified by them. To overcome this problem, organoleptic assessment (sensory score in tables 1 and 2) of virgin olive oil was proposed. This parameter was introduced in the 1980s by the International Olive Oil Council. And later, the European Commission included this parameter used to quantify the quality of olive oil. The sensory features depend on the phenolic concentration. The sensory parameters of this assessment are related with the aroma, flavor, pungency and bitterness of the virgin olive oil evaluated. In particular, the evaluation is based on the intensity of positive and negative (caused by human error) sensory descriptors between 0 and 10, Table 4. The final evaluation consists of the mean of all evaluations, carried out by a group of 8-12 experts in environmental conditions established by an Official methodology.

In addition to the aforementioned parameters, the concentration of other chemicals can quantify in more detail the quality of each category of olive oil, table 5. Some are included in the International Standards, such as the concentration of α-tocopherol, metals, insoluble impurities, *etc.* and others, which are not included in the International Standard such as polyphenols content, volatile compounds, pigments, contaminants, *etc.* can be found.

**Table 4. Positive and negative descriptors in the sensory assessments of virgin olive oil.**

| Positive | Negative |
|----------|----------|
| Fruity | Fusty |
| Green | Musty |
| Spicy | Muddy |
| Pungent | Winey |
| Bitter | Metallic |
| Over-ripe | Rancid |
| Fragrant | Burnt |
| Tropical | |
| Soft | |

**Table 5. Principal parameters to quantify the quality of olive oils**

| Parameter | Description | Analysis |
|-----------|-------------|----------|
| Free acidity (Free fatty acids) | Indicate the extension of the hydrolytic activities. | Titration of free fatty acids diluted in a mixture of solvents by an aqueous or ethanolic potassium hydroxide solution. |
| Peroxide value | Degree of oxidation, mainly occurring during storage (oxygen, light, storage time, temperature, *etc.*). | Iodometric procedure. |
| UV absorbance values at 232 and 270 nm ($K_{232}$, $K_{270}$ and $\Delta K$, equation 1) | Degree of olive oil oxidation and the apparition of new substances. | Determination UV absorbance at 232 and 270 nm of oil samples diluted in adequate solvents. |
| Sensory score | Determination of negative sensory perception. | Group composed of 8-12 expert taster evaluate the quality of the olive oil following the Panel test method. |

## 3. FACTORS WITH INFLUENCE ON OLIVE OIL QUALITY

Olive oil quality is mainly based on sensorial features and free acidity content and the aforementioned parameters. In comparison with other edible oils, the free acidity ranges between 0 and 2 % in the form of oleic acid, is of the least value. Due to its chemical composition (polar phenols and $\alpha$-tocopherol), olive oil is one of the most stable edible oils (Boskou, 2006). Nevertheless, the oxidization and hydrolytic processes have a notable influence on olive oil quality. Although these processes affect its organoleptic (flavor, taste, color, *etc.*) and nutritional characteristics, this point is not studied in enough depth in the current legislation.

### 3.1 Manufacturing Process

Focusing on the manufacturing process of virgin olive oil, its quality can be affected at every processing stage. The negative influences on the quality during various stages and how the manufacturing processes can be improved are presented here. The first and most important detail, obviously, consists of starting with good quality fruits, using the best equipment and manufacturing flow in proper operating conditions with good sanitation and good manufacturing practices.

### *Harvesting Of Olives*
Because the antioxidant quantity depends on the maturity of the olives, their harvesting should take place at the appropriate moment of ripeness (Gimeno *et al.*, 2002). In general, the ripened olives do not greatly affect their free acidity, peroxide index and spectrophotometric absorption $K_{232}$ and $K_{270}$. Nevertheless, Baccoury *et al.* found a noticeable decrease of oleic/linoleic acid ratio with the ripened olives (Baccoury *et al.*, 2008). This harvesting should be done with great care. Although harvesting by hand is the most expensive method (it causes minimum damage to the olives), it is one of the best options. Mechanical shaker harvesters are one of the cheapest methods but they will slightly damage the fruit, and as a result, the degradation rate of the olive is higher. The selection of the appropriate harvest method depends on various factors, but two main perspectives should be kept in mind; the quality required and economic considerations.

## Transporting Olives

In order to cause minimum damage to the olives, open bags or crates are the most appropriate method to transport them from the orchards to the factory. Given that good ventilation is present, harmful heating, and therefore, catabolic activity is prevented. In general, improper handling during transport of the olives can initiate the degradation by incipient enzymatic reactions, affecting the organoleptic characteristics and physicochemical parameters, and the resulting quality of the extra virgin olive oil produced is negatively affected (Boskou, 2006).

## Storage Of Olives

At this stage, cold storage of fruits can reduce the oxidative rate of the olive. It is necessary to take into account that storage temperatures of lower than 3 °C can be harmful to the fruit. At room temperature, an immediate processing (less than 10 hours after delivery) is the best option to attain the best possible quality of virgin olive oil. Pereira *et al.* found that the storage of olives in plastic containers at 5 °C for periods of up to 14 days negatively affect the extra virgin olive oil quality produced, i.e., making the acidity, resistance to oxidation, and total tocopherol content of the olive produced by this storage fruits poorer (Pereira *et al.*, 2002).

## Pre-Treatments of Olives

The olives must be cleaned to prevent the sensory characteristic deterioration of the olive oil. The cleaning process is carried out by a powerful air flow to remove leaves, stems, branches, dirt, *etc.* Then, the olives should only be washed using water when they are dirty or contaminated by pesticides. In general, they should not be washed with water because in doing so the moisture content is increased by about 5 %, and the extra virgin olive oil shows less fruitiness, bitterness and pungency characteristics.

## Crushing Olives

Although each type of mill presents different advantages and disadvantages, all types of mills used to make extra virgin olive oil can produce an excellent quality of oily juice. The influence of the type of mill used on the main characteristics of the olive oil produced (fruity, green, sweetness, pungent, and bitter) is shown in table 6. The temperature of the olive paste and the time required in the later malaxation process vary according to the type of mill used. Usually, the selection of the correct mill should be based on experience and scientific knowledge (Vossen, 2007).

**Table 6. The effect of different milling machines on the organoleptic characteristics of the olive oil.**

| Types of Mills | Fruity | Green | Sweetness | Pungent | Bitter |
|---|---|---|---|---|---|
| Hammer Mill | 🖐 | 🖐 | ☝ | 🖐 | 🖐 |
| Stone mill | ☝ | ☝ | 🖐 | ☝ | ☝ |
| Disc mill | ✌ | ✌ | ✌ | ✌ | ✌ |
| Pitter mill | ☝ | ☝ | 🖐 | ☝ | ☝ |

🖐 - Good.

✌ - No significant influence.

☝ - Bad.

## Malaxation of Olive Paste

In the malaxation process, the temperature and its duration are the most influential variables on the extraction capability of virgin olive oil from the olive paste (Kalua *et al.*, 2006; Amirante *et al.*, 2001). A temperature during the beating process higher than 30 °C could influence the amount of olive oil extracted from the paste and increases the polyphenol content and bitterness of virgin olive oil, but decreases the aromatic content. Beating the olive paste at 35 °C instead of at 27 °C, the oxidation resistance decreases by about 30 % (Amirante *et al.*, 2001). Longer malaxation (30-60 minutes) reduces the polyphenols, bitterness, stability and increases color and oxidation of the olive oil. In addition, in relation with the results obtained, mixing for more than 90 minutes can slightly influence the final quality of virgin olive oil. In general, due to enzymatic activity, longer blending periods gives the olive oil a pleasant aroma at the expense of stability because of a decrease in the antioxidant concentration. On the other hand, olive paste with high moisture values can have a negative influence on the olive oil extraction performance. Under some specific conditions, using talc can increase the extractability by up 5 %, because this mineral can absorb 15 times its weight in water (International Olive Council, 2008).

## Olive Oil Extraction Processes

In this phase, to prevent the fermentation process of the solids in the virgin olive oil, a rapid method to separate the oily phase from other compounds is required. Currently, several methods based on selective filtrations have been proposed to separate the oil from the olive paste, but these processes present lower rates of extracted oil than the continuous processes, two-phase and three phase and percolation methods. In the most widely used extraction process (pressure method, three phase, two phase and percolation methods), dirt problems appear and because of this the operator needs to keep track of the cleanliness of the oil (Vossen, 2007; Miled *et al.*, 2000). The cleaned olive oil produced using any method should contain less than 0.1 % of water or solids.

**Pressure method.** The main handicap to achieving the highest quality characteristics using this method is related with the aforementioned cleaning of the oil diaphragms. Fermentation processes could be triggered by leftover paste on the oil diaphragms, and then, the olive oil produced would be contaminated.

**The three-phase decanter** requires the addition of water to the system, which dilutes the water soluble components such as polyphenols, and these compounds, dissolved in waste water, are lost. As the produced virgin olive oil has lower natural antioxidant concentration, it presents an oxidative stability lower than is produced using other techniques (pressure, two-phase or percolation methods). Amirante *et al.* stated that the virgin olive oil extracted from de-stoned olive paste showed better resistance to oxidation and had a higher total of volatile compound concentration and the oil showed better taste qualities than those in olive extracted from whole paste. They also affirmed that resistance to oxidation increases with the total phenol concentration (Amirante *et al.*, 2006; Gimeno *et al.*, 2002).

**The two-phase decanter.** In most cases, this decanter does not require the addition of water, thus the manufacture of virgin olive oil using this method generates a lesser amount of wastewater. And as a much lesser amount of polyphenols can be dissolved in a smaller quantity of water (vegetable water, water present in olives), their concentration in the oily phase (virgin olive oil) is higher. This confers on the olive oil manufacture by the two-phase decanter a greater stability against oxidation for longer-term conservation than olive oil produced by three phase decanter or even by pressure methods. In addition, as opposed to pressure methods, both centrifugal processes provide extra virgin olive oil with a high concentration of chlorophyll.

**Percolation/centrifugation process.** Due to the polyphenol content in extra virgin olive oil, it presents similar quality characteristics to those produced by pressure methods (Boskou, 2006) and also, this oil presents a high oxidative stability.

Different quality indices of extra virgin olive oil manufacture by the aforementioned extractions process are shown in table 7. As can be seen, due to the water added in the three phase centrifugation, the difference is mainly focused on the concentration of phenolic compounds. Ranalli *et al.* compared virgin olive oils from percolation (first extraction) with the corresponding oils from centrifugation (second extraction). Resistance to autoxidation, sensory score, the tocopherol, phenol and aromatic content of percolated oil were the highest (Ranalli *et al.*, 1999).

**Table 7. Quality indices of virgin olive oil manufacture using pressing, three-phase and two-phase centrifugation and percolation.**

| Qualities features of olive oils manufactured by pressing, three-phase centrifugation and percolation (Di Giovacchino, 1996) | | | |
|---|---|---|---|
| Quality indices | Pressing | Three –phase | Percolation |
| Acidity (%) | 0.23 | 0.22 | 0.23 |
| Peroxide value (meq.O$_2$/kg) | 4.0 | 4.9 | 4.6 |
| Total polyphenol (mg mL$^{-1}$) | 158 | 121 | 157 |
| o-diphenols (mg mL$^{-1}$) | 100 | 61 | 99 |
| Chlorophyll (ppm) | 11.7 | 9.1 | 8.9 |
| K$_{232}$ | 1.93 | 2.01 | 2.03 |
| K$_{270}$ | 0.12 | 0.127 | 0.124 |
| Organoleptic rating | 6.9 | 7.0 | 7 |
| Qualities features of Cornicabra virgin olive oils manufactured by pressing, three-phase and two phase centrifugation (Salvador *et al.*, 2003) | | | |
| Quality indices | Pressing | Three –phase | Two–phase |
| Acidity (%) | 0.86 | 0.58 | 0.58 |
| Peroxide value (meq.O$_2$/kg) | 11.1 | 9.4 | 10.2 |
| K$_{232}$ | 1.653 | 1.616 | 1.619 |
| K$_{270}$ | 0.140 | 0.132 | 0.139 |
| Oxidative stability (h) | 46.3 | 57.2 | 65.8 |
| Total polyphenol (mg kg$^{-1}$) | 100 | 142 | 160 |
| o-diphenols (mg kg$^{-1}$) | - | 6.9 | 9.2 |
| α-Tocopherol (mg kg$^{-1}$) | 134 | 160 | 178 |
| Chlorophyll (ppm) | 11.4 | 8.6 | 11.4 |
| Carotenoids (mg kg$^{-1}$) | 6.8 | 6.5 | 7.6 |
| Intensity of bitterness | - | 1.6 | 2.0 |
| Organoleptic rating | 6.1 | 6.5 | 6.4 |

*Clarification*

As virgin olive oil at less than 12 °C may cause the crystallization of some saturated fatty acids, or if the environmental temperature is too warm, the olive oil can lose its flavor characteristics. The temperature of the process (sedimentation tanks or clarifier) is the most relevant variable in the quality of the virgin olive oil. Nevertheless, in all cases, the extra virgin olive oil can be sold without this sedimentation, but in this case, the user should be advised.

## Storage of Olive Oil

During the storage of extra virgin olive oil, due to enzymatic activity, degradation reactions such as oxidization and hydrolytic reactions can take place. Although olive oil has antioxidant compounds which prevent this type of decomposition reaction, as a consequence of inadequate storage conditions, virgin olive oil can suffer oxidization processes with a resulting decrease in quality. Because of this, the storage at factories and in homes must be adequate. Current legislation recommends the consumption of the extra virgin olive oil within a period of up to 18 months from the bottling date in order to obtain the maximum benefits from its physicochemical and organoleptic characteristics.

Regarding the influence of light on the stability of virgin olive oil throughout its storage, olive oil exposed to diffused daylight and artificial light attained maximum peroxide value in the second or third month of storage and decreased thereafter, while samples stored in the dark attained their maximum peroxide value in the sixth month of storage (Vekiari et al., 2007). Although light and climatologic conditions during the storage time affect the quality of olive oil, this influence varies depending on its chemical composition, i.e. good quality extra virgin olive oils resist the degradation better than poorer quality olive oils. The type of material used for the container also affects the stability of the oil. Containers made of stainless steel, tinplate, glass, clear PET bottle, PET bottle (covered with Aluminum foil), tetra-brik, etc. which protects this oily juice from oxygen and light are suitable alternatives to prevent its degradation. Méndez and collaborators found a gradual loss of quality during storage, especially in plastic and glass bottles. It is worthwhile to state that the best container for the commercialization of extra virgin olive oil is made of tin and tetra-brik (Méndez and Falque, 2007). Given the complexity of estimating the quality of packed virgin olive oil, mathematical models have been proposed to predict this (Coutelieris and Kanavouras, 2006). In addition, during the bottling process or storage of the olive oil, some manufacturers replace the oxygen which remains in the container by nitrogen as an inert gas because in these conditions the degradation reaction is notably reduced. Because of this, when the container is periodically opened, the oxygen from the container is renewed and the oxidation reaction in the olive oil is much faster. In addition, because extra virgin olive oil contains pheophytin compounds which are endogenous photosensitizers (similar to chlorophyll), the oxidative process of the oil is increased in the presence of light. In open containers, the oxidation using atmospheric oxygen and photosensitized processes work in parallel. Because of this, protection from oxygen and light inhibits and can even prevent this oxidation process.

The importance of this stage regarding consumption is clear and therefore storage conditions should be optimized. Mathematical models based on the degradation reaction kinetics and/or indexes related with the resistance to oxidative reactions have been proposed. The degradation and characteristics of the oil can be estimated, and then, the storage conditions can be optimized by these mathematical models (Boskou, 2006).

In general, the production of the highest quality extra virgin olive oil requires the optimum application of three stages, *viz.* firstly, producing an olive fruit with superior qualities, then transferring these positive characteristics to the olive juice produced, and finally, maintaining the nutritional benefits throughout the commercial life of the oil (Kalua *et al.*, 2007). This should be the most important objective of every producer in the olive sector. In addition, as the chemical composition of olives vary considerably from one year to the next (Salvador *et al.*, 2003), the operational conditions (temperature, malaxation time, water added, *etc.*) should be suitably optimized depending on the circumstances (Servili *et al.*, 2003).

# 4. CONCLUSIONS

In the food sector and especially in the field of relatively expensive foods, the maintenance of a level of high quality is required. Extra virgin olive oil is a clear example of this. Currently, different chemical and mathematical methods based on chemical composition or kinetic aspects are being proposed to prevent not only economic fraud but also to protect consumers from any potential health hazards.

With this aim, the two most important classifications of olive oils in the world have been summarized here (one provided by United States Department of Agriculture and the other made available by the International Olive Oil Council, Codex Alimentarius and the European Commission). Given that most of the members of the International Olive Oil Council are from the Mediterranean basin, its classification is the most used in Europe.

In order to produce the highest quality olive oil, the cultivation, harvesting, transport of the olives, the manufacturing process and the final storage at the factories and in the home should be carried out with care. The handling of olives, operation temperature, light, chemical composition, *etc.* are aspects with notable influence on the degradation process (mainly oxidative and hydrolyte reactions) of olive oil. These must be controlled to achieve a maximum return for the manufacturer and to give consumers the best possible quality of olive oil. A summary with interesting scientific references have been provided here.

# REFERENCES

Amirante, P., Clodoveo, M. L., Dugo, G., Leone, A. & Tamborrino, A. (2006). Advance technology in virgin olive oil production from traditional and de-stoned pastes: Influence of the introduction of a heat exchanger on oil quality. *Food Chemistry, 98*, 797–805.

Amirante, R., Cioni, E, Montel, G. L. & Montel Pascualone, A. (2001). Influence of mixing and extraction parameters on virgin olive oil quality. *Grasas y Aceites, 52*, 198-201.

Angiuli, M., Ferrari, C., Lepori, L., Matteoli, E., Salvetti, G., Tombari, E., Banti, A. & Minnaja, N. (2006). On testing quality and traceability of virgin olive oil by calorimetry. *Journal of Thermal Analysis and Calorimetry, 84*, 105–112.

Baccouri, O., Guerfel, M., Baccouri, B., Cerretani, L., Bendini, A., Lercker, G., Zarrouk, M. & Miled, D. D. B. (2008). Chemical composition and oxidative stability of Tunisian monovarietal virgin olive oils with regard to fruit ripening. *Food Chemistry, 109*, 743–754.

Bilancia, M. T., Caponio, F., Sikorska, E., Pasqualone, A. & Summo, C. (2007). Correlation of triacylglycerol oligopolymers and oxidized triacylglycerols to quality parameters in extra virgin olive oil during storage. *Food Research International, 40*, 855–861.

Boskou, D. Olive oil: Chemistry and technology. 2nd Edition. Champaign, Illinois: AOCS PRESS; 2006.

Ciafardine, G. & Zullo, B. A. (2002). Microbiological activity in stored olive oil. *International Journal of Food Microbiology, 75*, 111-118.

Ciafardine, G. & Zullo, B. A. (2002a). Survival of micro-organims in extra virgin olive oil during storage. *Food Microbiology, 19*, 105-109.

Coutelieris, F. A. & Kanavouras, A. (2006). Experimental and theoretical investigation of packaged olive oil: Development of a quality indicator based on mathematical predictions. *Journal of Food Engineering, 73*, 85–92.

Di Giovacchio, L., Influence of extraction system on olive oil quality. Olive oil, Chemistry and Technology, AOCS Press, Champaign Illinois, 12-51, 1996.

Dourtoglou, V. G., Dourtoglou, T., Antonopoulos, A., Stefanou, E., Lalas, S. & Poulos, C. (2003). Detection of olive oil adulteration using principal component analysis Applied on Total and Regio FA Content. *Journal of American Oil Chemists Society, 80*, 203-208.

Duran, R. M. (1990). Relationship between the composition and ripening of the olive and the quality of the oil. *Acta Horticulturae, 286*, 441–451.

Fonseca, A. M., Biscaya, J. L., Aires-de-Sousa, J. & Lobo, A. M. (2006). Geographical classification of crude oils by Kohonen selforganizing maps. *Analytica Chimica Acta, 556*, 374–382.

Gimeno, E., Castellote, A. I., Lamuela-Raventós, R. M., De la Torre, M. C. & López-Sabater, M. C. (2002). The effects of harvest and extraction methods on the antioxidant content (phenolics, α-tocopherol, and β carotene) in virgin olive oil. *Food chemistry, 78*, 207-211.

Gould, W. A. Total quality management for the food industries. Baltimore: CTI Publications Inc.; 1992.

Grigoriadou, D. & Tsimidou, M. (2006). Quality control and storage studies of virgin olive oil. Exploiting UV spectrophotometry potential. *European Journal of Lipid Science and Technology, 108*, 61-69.

International Olive Council. 2008. (Last visited August 2009) URL http://www.internationaloliveoil.org.

Kalua, C. M., Allen, M. S., Bedgood Jr, D. R., Bishop, A. G., Prenzler, P. D. & Robards, K. (2007). Olive oil volatile compounds, flavour development and quality: A critical review. *Food Chemistry, 100*, 273–286.

Kalua, C. M., Bedgood, D. R., Bishop, A. G. & Prenzler, P. D. (2006). Changes in Volatile and Phenolic Compounds with Malaxation Time and Temperature during Virgin Olive Oil Production. *Journal of Agricultural and Food Chemistry, 54*, 7641-7651.

Lee, D. S., Noh, B. S., Bae, S. Y. & Kim, K. (1998). Characterization of fatty acids composition in vegetable oils by gas chromatography and chemometrics. *Analytica Chimica Acta, 358*, 163-175.

Lerma-García, M. J., Herrero-Martínez, J. M., Ramis-Ramos, G. & Simó-Alfonso, E. F. (2008). Evaluation of the quality of olive oil using fatty acid profiles by direct infusion electrospray ionization mass spectrometry. *Food Chemistry, 107*, 1307–1313.

Marini, F., Magrì, A. L., Bucci, R. & Magrì, A. D. (2007). Use of different artificial neural networks to resolve binary blends of monocultivar Italian olive oils. *Analytica Chimica Acta, 599*, 232–240.

Méndez, A. I. & Falqué, E. (2007). Effect of storage time and container type on the quality of extra-virgin olive oil. *Food Control, 18*, 521–529.

Mildner-Szkudlarz, S. & Jelen, H. H. (2008). The potential of different techniques for volatile compounds analysis coupled with PCA for the detection of the adulteration of olive oil with hazelnut oil. *Food Chemistry, 110*, 751–761.

Miled, D. D. B., Smaoui, A., Zarrouk, M. & Chérif, A. (2000). Do extraction procedures affect olive oil quality and stability?. *Biochemical Society Transactions, 28*, 929-933.

Peña, F., Cárdenas, S., Gallego, M. & Valcárcel, M. (2005). Direct olive oil authentication: Detection of adulteration of olive oil with hazelnut oil by direct coupling of headspace and mass spectrometry, and multivariate regression techniques. *Journal of Chromatography A, 1074*, 215–221.

Pereira, J. A., Casal, S., Bento, A. & Oliveira, M. B. P. P. (2002). Influence of Olive Storage Period on Oil Quality of Three Portuguese Cultivars of Olea europea, Cobrancuosa, Madural, and Verdeal Transmontana. *Journal of Agricultural and Food Chemistry, 50*, 6335-6340

Ranalli, A., Ferrante, M. L., De Mattia, G. & Costantini, N. (1999). Analytical evaluation of virgin olive oil of first and second extraction. *Journal of Agricultural and Food Chemistry, 47*, 417-424.

Salvador, M. D., Aranda, F., Gómez-Alonso, S. & Fregapane, G. (2003). Influence of extraction system, production year and area on Cornicabra virgin olive oil: a study of five crop seasons. *Food Chemistry, 80*, 359-366.

Sayago, A., García-González, D. L., Morales, M. T. & Aparicio, R. (2007). Detection of the presence of refined hazelnut oil in refined olive oil by fluorescence spectroscopy. *Journal of Agricultural and Food Chemistry, 55*, 2068-2071.

Servili, M., Selvaggini, R., Taticchi, A., Esposto, S. & Montedoro, G. (2003). Volatile Compounds and Phenolic Composition of Virgin Olive Oil: Optimization of Temperature and Time of Exposure of Olive Pastes to Air Contact during the Mechanical Extraction Process. *Journal of Agricultural and Food Chemistry, 51*, 7980-7988.

Torrecilla, J. S., Rojo, E., Oliet, M., Domínguez, J. C. & Rodríguez, F. (2009). Self-organizing maps and learning vector quantization networks as tools to identify vegetable oils. *Journal of Agricultural and Food Chemistry, 57*, 2763-2769.

United States Department of Agriculture, United States Standards for Grades of Olive Oil, (Last visited August 2009) URL http://www.agmrc.org/media/cms/oliveoil_08D51C4E3F9DA.pdf.

United States Department of Agriculture, United States Standards for Grades of Olive Oil. (Last visited August 2009) URL http://www.ams.usda.gov/AMSv1.0/getfile?dDocName=STELPRDC506978 0.

Vekiari, S. A., Papadopoulou, P. & Kiritsakis, A. (2007). Effects of processing methods and commercial storage conditions on the extra virgin olive oil quality indexes. *Grasas y aceites, 58*, 237-242.

Vossen, P. (2007). Olive Oil: History, Production, and Characteristics of the World's Classic Oils. *Hortscience, 42*, 1093-1100.

*Chapter 6*

# INDUSTRIAL METHODS TO PRODUCE REFINED OLIVE OIL

## ABSTRACT

In the olive oil field, there are three broad groups of oils, viz. the edible oily juice of the olive which does not require a chemical process, non edible lampante virgin olive oil and olive pomace oil. Once the highest grades of virgin olive oil have been produced (extra virgin olive oil and virgin olive oil) from the olive, lampante olive oil remains. And finally, the lowest grade, olive pomace oil is produced from the pomace. Here, the main stages in the conventional method to produce refined olive oil (neutralization, bleaching, deodorization and winterization processes), and its chemicals, which can appear or be removed in each of them have been shown. Non conventional methods based on supercritical extractions, microfiltration and "sparging", used to produce olive oil is also presented here. Finally, the hydrogenation, interesterification and glycerolysis of refined olive oil have been mentioned to explain the manufacture of other compounds such as margarine, mayonnaise, salad dressing, *etc.* which add to the economic return of the oleic sector.

**Keywords**: Refined olive oil; Lampante olive oil; Neutralization; Bleaching process; Deodorization; Winterization.

# 1. INTRODUCTION

Once all possible edible categories have been extracted from the olives, waste (aqueous phase and/or olive pomace, whose management will be studied in section 4) and non edible lampante virgin olive oil remain (see Chapter 5). Although this oil is still more expensive than other seed oils, due to its chemical composition and sensory features, this oil is not edible, and therefore, the European Commission does not allow its commercialization. At this point, as with other non edible oils, chemical treatments or refining processes are required to maximize the value of this sub standard oil.

During the 2006-2007 and 2007-2008, depending on the harvest region, the price of refined olive oils was approximately constant in the range of 2.3 and 2.8 €·kg$^{-1}$. Nevertheless, during the 2008/09 season, its price has decreased to 1.6 and 2.1 €·kg$^{-1}$. Regarding refined olive pomace oil, during the 2006-2007 and 2007-2008 their prices increased from 1.4 to 1.7 €·kg$^{-1}$. During the 2008/09 season, as with virgin olive oil, the price of olive pomace oil has dropped to around 1 €·kg$^{-1}$ (International Olive Council, 2009). This fall in price represents a reduction in income of 27 % in the refined olive oil trade and 32 % in the olive pomace oil trade. It is certain that these falls in three of the most important trades of the olive oil sector will negatively affect the development of the olive oil sector and probably could affect the number of people employed (more than 800,000 in Europe encompassing approximately 600 companies in the European Union).

Usually, this type of process consists of four stages, neutralizing, bleaching, deodorizing and winterization. Following the European Commission's rules, once the refined olive oil has been produced (the bland fatty material has been obtained), in order to be commercialized, this refined olive oil must be blended with edible virgin olive oils (see Chapter 5). By this blending, similar organoleptic features to virgin olive oil are given to the refined olive oil.

# 2. MANUFACTURE OF REFINED OLIVE OIL

The ideal refining process should consist of maintaining the concentration and structure of valuable chemicals and the removal of hazardous chemical concentrations as much as possible. Depending on free fatty acids (commonly called acidity), UV absorbance values, peroxide values and sensory score, a given olive oil can be considered as edible in line with EC regulations. Alternatively, it must be chemically treated, and later blended with virgin olive oil to give it the characteristic flavor of olive oil. Throughout the refining process, major or minor constituents like free fatty acids, oxidization products, hydrocarbons, *etc.* are removed, increasing the stability of the produced oil. Over 50 % of the oil produced in the Mediterranean area is of such poor quality that it must be refined to produce an edible product. As will be explained later, refined olive oil can be made by conventional or non conventional methods.

**Table 1. Refining stages of edible oils and major impurities removed or reduced.**

| Refining stage | Major impurities removed or reduced |
|---|---|
| Degumming | Phospholipids, trace metals, pigments, carbohydrates, proteins. |
| Neutralization | Fatty acids, phospholipids, pigments, trace metals, sulfur, insoluble matter. |
| Bleaching | Pigments, oxidation products, trace metals, traces of soap, *etc.* |
| Filtration | Spent bleaching earth. |
| Deodorization | Fatty acids, mono and diacylglyderols, oxidization products, pigments, decomposed products, pesticides, *etc.* |
| Winterization | Waxes, saturated fractions of the oil. |

## 2.1. Conventional Method

Generally, the refining of lampante olive oil made by the conventional method is carried out at relative high temperatures, where some necessary natural compounds are regretfully lost and non required chemicals are generated. The refining process is dependent on the quality of the lampante olive oil to be treated, and therefore processes (and operating conditions) vary according to quality. In general, the conventional method consists of five stages: (*i*) The neutralization process is required with the main aim to reduce acidity; (*ii*) To produce edible olive oil, the bleaching process is used to reduce the products generated during the oxidization process of the olives; (*iii*) the deodorization process is carried out to remove the aldehyde, fatty acids and hazardous components. At this stage, low acidity and peroxide concentration values can be assumed; (*iv*) The winterization process is used to remove waxes and high melting point triacylglycerols present in the lampante virgin olive oil; (*v*) Finally, to produce edible refined olive oil which complies with the regulations to be classified as "olive oil" (see Chapter 5), the refined olive oil produced is blended with virgin olive oil. In table 1, the main stages of the refining of edible oils and the major impurities removed or reduced have been summarized. Currently, due to the fact that modern equipment can measure trace amounts, researchers have found that during the refining process of the lampante virgin olive oil, small quantities of chemicals are generated. This generation is used to quantify its quality and detect possible adulteration of refined olive oil (Boskou, 2006). A more detailed description is given below.

## Neutralization Process

The acidity grade of the lampante virgin olive oil conditions the neutralization process. Usually, the free fatty acids are removed by an alkali such as sodium hydroxide. Also in the presence of water, compounds such as mucilage and resinous substances become insoluble. All these chemicals can be separated simultaneously by filtering the oil. The adequate removal of the latter compounds facilitates the contact between activated carbon, which will be used in the posterior bleaching process of the neutralized olive oil. In this way, the bleaching rate increases. During this stage, in addition to the fatty acids, pigments and phospholipids are also removed. Throughout the neutralization process, about 15 % of total sterols are lost (Boscou, 2006).

## Bleaching Process

At this stage, in order to give a clear color to the refined olive oil, colored particles in the oil are eliminated. Other substances such as oxidative materials, trace metals, chlorophyll, carotenoids and residual fatty acids are also removed. This is achieved in two steps. The chemicals to be removed are absorbed in diatomaceous earth, synthetic silicas, activated earth or carbons, and then, these chemicals are separated by filtration. In addition, depending on the circumstances, the oils retained in the absorbents can be separated by extraction, using with solvents and by distillation.

During the refining process of the lampante olive oil, and in particular throughout the bleaching process, due to the action of other parallel chemical reactions, new chemicals can appear or existing ones can be removed. Some of them are shown below:

- As a consequence of using activated earths to carry out the bleaching process, the conjugation of part of the double bounds of the bi- and tri-unsaturated compounds takes place. In addition, dienes and trines are also formed. The presence of these chemicals can be detected by UV spectroscopy at absorbance values around 232 and 270 nm wavelength.
- Although the *trans* isomers appear in the deodorization process, throughout the bleaching process, the generation of these compounds are incipiently formed.

- From the point of view of hydrocarbons, during this stage, the squalene, which is 50 % of the total un-saponificable matter in the virgin olive oil, is notably reduced.
- In this stage, the sterols are also slightly reduced.

## *Deodorization Process*

The main objective of this stage is to remove all chemicals which are responsible for non adequate organoleptic features. During this process, volatile products, oxidation products, carotenoids, pesticides residues, sterols, tocopherols, hydrocarbon, unpleasant taste and odor and a large amount of free fatty acids are removed. The deodorization process is normally performed by the distillation process with steam under vacuum conditions. The steam is passed through the oil under vacuum conditions (400 – 1333 Pa) and at temperatures ranging between 200 and 250 °C. Temperatures higher than 250 °C may cause the formation of *trans* fatty acids.

Throughout the deodorization process, as a consequence of the parallel chemical reactions, new chemicals appear or disappear, changing the composition of the olive oil. Some examples are shown below:

- If a high temperature is used during the deodorization process, the modification of fatty acids structure, and then, the formation of *trans* isomers is notably favored and therefore, a higher quantity of these chemicals is produced. The classification of the different oil qualities depends on their chemical concentration and is regulated by European Commission Regulations (EC 640/2008), table 2. As can be seen, the limits vary depending on the chemical treatments necessary to make them edible. For instance, olive pomace oil, which is the poorest quality edible olive oil, presents the highest values of isomers concentrations ($\leq 0.40$ and $\leq 0.35$ for sum of isomers *trans* oleic acid and sum of isomers of *trans* linoleic and *trans* linolenic acids, respectively). In addition, the concentrations shown in table 2 guarantee the authenticity and sound manufacturing practice.
- The squalene, whose concentration was reduced in the bleaching process, can be recovered here from the sludge of deodorization.
- During this stage and the previous one, about 10 % of total sterols are reduced.
- Given that greater losses of α-tocopherol are inevitable, mainly during the deodorization process, the International Olive oil Council and Codex

Alimentarius permit the addition of this chemical up to 200 mg·kg⁻¹ to restore the losses of natural α-tocopherol during the refining process (Rabascal and Boatella, 1987; Beare-Rogers *et al.*, 2001).

- Longer deodorization processes at high temperatures produce an increase of saturated fatty acids in position 2 of the glycerol molecules. This concentration is used to detect the concentration of anomalous esterified oils and their adulterations.

**Table 2. Depending on the olive oil category, percentage of the geometrical isomers (*trans* oleic, *trans* linoleic, and *trans* linolenic acids). (EC 640/2008)**

| Categories of olive oils | | Sum of isomers *trans* oleic (%) | Sum of isomers of *trans* linoleic and *trans* linolenic (%) |
|---|---|---|---|
| Edible | Extra virgin olive oil | ≤ 0.05 | ≤ 0.05 |
| | Virgin olive oil | ≤ 0.05 | ≤ 0.05 |
| | Olive oil | ≤ 0.20 | ≤ 0.30 |
| | Olive pomace oil | ≤ 0.40 | ≤ 0.35 |
| Non edible | Lampante olive oil | ≤ 0.10 | ≤ 0.10 |
| | Refined olive oil | ≤ 0.20 | ≤ 0.30 |
| | Refined olive-pomace oil | ≤ 0.40 | ≤ 0.35 |

*Winterization Process*

This process is also referred to as fractionation. As the waxes, saturated fraction of the olive oil and high melting point triacylglycerols are removed from the lampante olive oil by means of this process, the treated olive oil is made more resistant to sedimentation and clouding effects. This process consists of two steps, firstly the olive oil is kept in crystallization tanks at about 5 °C for between 24 to 36 hours, and finally, the oil is filtered. Using this method, the waxes are retained in the solid phase.

## 2.2. Non Conventional Methods

Trying to reach the aforementioned ideal refining process of the lampante olive oil, different processes have been proposed. Although these methods are not well known on an industrial scale, different approaches are proposed here as examples.

## Refining by Supercritical Extraction

This process consists of a supercritical carbon dioxide extraction plant operating in a continuous countercurrent mode. By this method, the refined olive oil is produced without the negative aspects of conventional methods (in relation with the appearance of non desirable chemicals). And although the cost is higher, the results stated by Bondioli *et al.* and in the Ganguli and collaborators' patent demonstrate the possibility of fractionating components contained in the initial olive oil even if there are only trace elements present (Bondioli *et al.*, 1992; Ganguli *et al.*, 1997; Boskou, 2006).

## Microfiltration Process

This process consists of removing minor and major constituents by means of deacidification by sodium hydroxide neutralization and membrane microfiltration processes. The neutralization of the free fatty acids with an appropriate soda concentration allows the formation of submicronic particles which are subsequently removed by a microfiltration technique. The filtered lampante olive oils show very low residual acidity as well as soap and water contents. In addition, the passage from the laboratory scale (1 kg) to a 50 kg unit shows the practical possibility of scaling up without meeting any particular problems or loss of efficiency (Hafidi *et al.*, 2005; Hafidi *et al.*, 2004).

## Mild Purification Process

By this process, refined olive oil with a high concentration of polyphenols and adequate flavor features is achieved. Fatty acids and unpleasant odorous substances are removed by means of "sparging" with an inert gas. In general, this process consists of heating the olive oil at temperatures ranging between 120 – 250 °C, under vacuum conditions between 2 - 5 mbar. In addition, steam is input into the bottom of the olive oil container, using preferably 1 - 15 $m^3$ steam per kg of olive oil. This water is converted into bubbles and these go up through the oil carrying with it water soluble compounds. Although at 250 °C one hour sparging has appeared to be a suitable process time, when refining takes place at 120 °C, the time should be prolonged to 5 hours. This procedure was defined by Van Buuren and collaborators and this has been patented as a procedure to produce refined olive oil (Buuren *et al.*, 2002).

## 3. APPLICATIONS OF REFINED OLIVE OIL

In addition to the manufacture of low grade olive oils, many other applications can be found for them. In particular, once the refined olive oil has been produced, in order to produce other products with higher added value and contribute to increase the economic return, other chemical processes such as hydrogenation or interesterification of this oil can be performed. For instance, the hydrogenated olive oil can be used to manufacture margarine, cooking fats, *etc.* As examples, three different chemical processes are shown below.

### *Hydrogenation of the Refined Olive Oil.*
Throughout the hydrogenation carried out under conditions which favor isomerization, dienoic fatty acids and geometrical and positional isomers are generated and contained in the final product. Although only small quantities of refined olive oil are used for this purpose, margarines and cooking fats can be made by this process (Boskou, 2006).

### *Interesterification of Refined Olive Oil*
By this process, in comparison with the aforementioned hydrogenation process, oily product with less *trans*-plastic fats and a higher concentration of polyunsaturated fatty acids (70-30) are produced. By combining hydrogenated palm oil and interestificated olive oil, margarine with the added advantage of presenting a higher concentration of monounsaturated fatty acids can be made (Alpalsan and Karaali, 1998). By the combination of beef oil and interestificated olive oil, frankfurters can be manufactured. In this case, a better ratio of unsaturated to saturated hydrocarbon is achieved (Vural *et al.*, 2004).

### *Glycerolysis of Refined Olive Oil*
Salad dressing and olive oil mayonnaise with similar rheological characteristics to conventional ones, but with a lesser number of calories can be produced by the glycerolysis and enzymatic *trans* esterification of refined olive oil (Tynek and Ledochowska, 2005; Fomuso *et al.*, 2001).

## 4. CONCLUSIONS

There are different edible olive oil categories, ranging from the pure oily juice of fruits to olive oils which come from the extraction by solvents of olive pomace oil. In all cases, as the progenitor is a fruit, the olive, by comparison with other seed oils, all low grade olive oils presents better organoleptic characteristics and are also healthier.

In the light of the chemical and organoleptic characteristics of the remaining lampante virgin olive oil, once the extra virgin olive oil and virgin olive oil have been extracted, the remaining oil is not edible and therefore, it cannot be commercialized. And so, to produce an edible olive oil, physicochemical processes are required. This remaining olive oil is called lampante virgin olive oil and it is the prime material in conventional and non conventional processes in the production of refined olive oil.

The conventional process has been presented here, and consists of neutralization, bleaching, deodorization and winterization processes. In addition, non conventional processes to produce refined olive oil based on supercritical extractions, microfiltration and "sparging" are also presented. As a final stage, common to both methods, there is the blending of this refined olive oil with virgin olive oil (this is fixed by European Commission's rules). Finally, the hydrogenation, interesterification and glicerolysis of the refined olive oil have been presented in brief to explain the manufacture of other compounds such as margarine, mayonnaise, salad dressing, *etc.* which increase the economic return in the olive oil sector.

## REFERENCES

Alpaslan, M. & Karaali, A. (1998). The interesterification-induced changes in olive and palm and oil blends. *Food Chemistry, 61*, 301-305.

Beare-Rogers, J., Dieffenbacher, A., Holm, J. V. (2001). Lexicon of lipid nutrition. *Pure and Applied Chemistry, 73*, 4, 685–744.

Bondioli, P., Mariani, C., Lanzani, A., Fedeli, E., Mossa, A. & Muller, A. (1992). Lampante olive oil refining with supercritical carbon dioxide. *Journal of the American Oil Chemists' Society, 69*, 477-480.

Boskou, D. Olive oil: Chemistry and technology. 2nd Edition. Champaign, Illinois: AOCS PRESS; 2006.

Buuren, J. V., Ganguli, K. L. & Putte, K. P. (2002). Olive oil containing food composition. European Patent EP1042435.

Fomuso, L. B., Corredig, M. & Akoh, C. C. (2001). A comparative study of mayonnaise and Italian dressing prepared with lipase-catalyzed transesterified olive oil and caprylic acid. *Journal of the American Oil Chemists' Society, 78*, 771-774.

Ganguli, K. L., Van, I. A. R., Michaelides, G., Polman, R. G. & Van, P. K. P. A. M. (1997). Olive oil blends European Patent EP0600539.

Hafidi, A., Pioch, D. & Ajana, H. (2005). Soft purification of lampante olive oil by microfiltration. *Food Chemistry, 92*, 17-22.

Hafidi, A., Pioch, D., Teyssier, M. L. & Ajana, H. (2004). Influence of oil conditioning on the permeate flux and cake properties during microfiltration of lampante olive oil. *European Journal of Lipid Science and Technology, 106*, 152–159.

International Olive oil Council. 2009. (Last visited August 2009) URL http://www.internationaloliveoil.org.

Rabascal, N. H. & Boatella, R. (1987). Variaciones del contenido en tocoferoles y tocotrienoles durante los procesos de obtención, refinación e hidrogenación de aceites comestibles. *Aceites y grasas, 38*, 145-150, 1987.

Tynek, M. & Ledochowska, E. (2005). Structured triacylglycerols containing behenic acid: Preparation and properties. *Journal of Food Lipids, 12*, 77-82.

Vural, H., Javidipour, I. & Ozbas, O. (2004). Effects of interesterified vegetable oils and sugar beet fiber on the quality of frankfurters. *Meat Science, 67*, 65-72.

# SECTION 4. WASTE FROM THE OLIVE OIL SECTOR

*Chapter 7*

# WASTE AND ITS MANAGEMENT IN THE OLIVE SECTOR

## ABSTRACT

The advantages related with human health and the highly palatable characteristics of extra virgin olive oil and table olives are clear. However, the environmental impact of the wastes generated during their manufacture cannot be forgotten. Virgin olive oil manufacture generates annually an amount of waste equivalent to that produced by 22 million people. In addition, the residues generated in the manufacture of table olives should be also into account. During the last decades, regretfully, despite many research groups having studied in depth the best method to treat these residues, the ideal treatment has still not been found. However, there are some treatments which are capable of reducing notably the environmental impact (detoxification of residues, separation process, *etc.*) and some of them can even convert these harmful wastes into material which is reusable in other sectors (land application, compost, energy, manufacture of synthesis gas, *etc.*). How the environment impact of residual waste can be reduced by the application of different techniques are given in this Chapter.

**Keywords:** Olive oil mill waste water; Olive pomace; Detoxification processes; Environmental impact.

# 1. INTRODUCTION

The advantages to health and the pleasant organoleptic characteristics of the main products of the olive sector such as extra virgin olive oil and table olives have been known for millenniums, and their consumption continues to increase throughout the world. In Europe, 800,000 people are employed in this sector and more than $4\cdot10^6$ hectares are used to satisfy this growing trend in consumption. This sector is the second most important agro-food sector in Europe.

Throughout the world, many research groups are trying to improve and optimize the manufacturing processes between care of the olive tree and its final waste (Viesca et al., 2007). Although the production of extra virgin olive oil or table olives depends on the variety, climatic and cultivar conditions among other factors, these are produced using similar methods. One of the main disadvantages of this manufacturing process is the high environmental impact of the wastes generated. In an attempt to lessen this impact, the management of this waste is being researched widely throughout the world (Arvanitoyannis et al., 2007). Here, this subject will be studied in two principal groups viz. wastes generated during the olive oil process and table oil manufacture.

# 2. WASTE GENERATED DURING VIRGIN OLIVE OIL PRODUCTION

During the manufacturing processes of virgin olive oil, three main residual products are generated viz. twigs and leaves, olive pomace and olive oil mill wastewater, figure 1 (vide infra). Although leaves and twigs are not great in volumes, every year a single olive tree can generate approximately 25 kg of these wastes. Given that they have no direct commercial value, they are used as animal nutrition, landfill or to increase the porosity in the composting process. In addition, due to 90 mg per gram of dry leaves being oleuropein, which presents antioxidant and anti-inflammatory properties, a high economic return can be obtained by extracting it from the leaves. Besides, other natural antioxidants such as hydroxytyrosol, which has become commercially available for research purposes and costs in a range of $US 1,000-2,000 per gram can be extracted (Obied et al., 2005; Le Tutour et al., 1992; Baldioli et al., 1996). There are patents where methods to extract these types of components are explained (Pizzichini and Russo, 2005; Villanova et al., 2006; Marco et al., 2007).

Figure 1. Olive pomace storage 1996/97 season (Oleícola El tejar Nuestra Señora de Araceli Sdad. Coop. Ltda., Córdoba, Spain).

The biggest problem in the virgin olive oil manufacture process is the olive pomace and olive oil mill waste water production, which are high in volume and have a high pollution impact (Kapellakis *et al.*, 2008; Torrecilla *et al.*, 2006). Olive oil mill waste water is considered one of the major pollutants and causes great problems in olive tree cultivation areas in many European countries.

During 2008, more than of $3 \cdot 10^7$ m$^3$ of olive oil waste was generated between early November and late February. In the Mediterranean area, the polluting characteristics and the short harvesting period are two of the principal environmental problems. To realize the importance of its environmental impact, this is equivalent to the pollution produced by more than 22 million people per year (Isidori *et al.*, 2005). Chemically, olive oil mill waste water, also called black water, is a toxic effluent. It presents a pH range between 4 and 6, Biological Oxygen Demand in 5 days (BOD$_5$) of $20 - 55$ g L$^{-1}$ and a chemical oxygen demand (COD) of $20 - 35$ g L$^{-1}$. On the other hand, the main characteristic of the olive pomace is shown in table 1.

The quantity of olive oil waste produced depends on the manufacturing process used (see Chapter 4). Table 2 shows that for every 1,000 kg of olives, in pressure process, 1,000 kg, in three-phase processes 1,500-1,800 kg and in two-phase (also called ecological system) processes 800-950 kg of olive waste are produced (Kapellakis *et al.*, 2008). From these quantities, 600, 1,000-1,200 and 0 kg of olive oil mill waste water are produced, respectively. In the two-phase method, the water is mainly contained in the solid waste (olive pomace). The two-phase method produces an olive pomace with a moisture content of around 63 wet basis-% (Aragón and Palancar, 2001; Torrecilla et al., 2006).

**Table 1. Physical and chemical features of olive pomace on dry bases (Torrecilla et al., 2006).**

| Parameter | Values |
|---|---|
| Low heating value (MJ Kg$^{-1}$) | 18 |
| Carbon (%) | 38-45 |
| Oxygen (%) | 20-34 |
| Hydrogen (%) | 4-5.5 |
| Nitrogen (%) | 0.8-4 |
| Sulfur (%) | 0.01-0.04 |
| Phosphor (%) | 0.25 |
| Potassium (%) | 1.8 |
| Polyphenols (ppm) | 23,000 |
| Sugars (%) | 4.8 |
| Ashes (%) | 4-12 |

**Table 2. Comparative data for olive oil extraction processes (Aragón and Palancar, 2001).**

| Process | Input | Input amount | Output | Output amount |
|---|---|---|---|---|
| Pressing | Olives | 1000 Kg | Oil | 200 kg |
| | Washing Water | 0.1-0.12 m³ | Solid Waste (25 % water + 6 % oil) | 400 kg |
| | Energy | 40-63 kWh | Waste water (88 % water) | 600 kg |
| Three-phase | Olives | 1000 kg | Oil | 200 kg |
| | Washing Water | 0.1-0.12 m³ | Solid Waste (50 % water + 4 % oil) | 500-600 kg |
| | Fresh water for decanter | 0.5-1 m³ | Waste water (94 % water + 1 % oil) | 1,000-1,200 kg |
| | Water to clean the impure oil | 10 Kg | | |
| | Energy | 90-117 kWh | | |
| Two-phase | Olives | 1,000 kg | Oil | 200 kg |
| | Washing Water | 0.1-0.12 m³ | Solid Waste (60 % water + 3 % oil) | 800-950 kg |
| | Energy | < 90-117 kWh | | |

In the European Community, olive oil production is carried out mainly by methods based on centrifugation, *i.e.*, three-phase or two-phase techniques. In particular, in Italy and Greece olive oil manufacture is mainly carried out by the three phase method, whereas in Spain olive oil is extracted by the two-phase method (Aragón and Palancar, 2001). Although both methods are similar, their wastes are notably different, and therefore, their correct management is also different.

## 2.1 Olives' Leaves and Pits

### Phenolic Compounds

Currently, there is an increased interest in *bio*active compounds, natural products, food additives, antioxidant compounds, *etc.* which are high added value compounds. These compounds can be found in the leaves of olive trees. Although some phenolic compounds are present in higher concentration in the olives, these compounds are present in significant amounts in the olive tree leaves capable of being extracted. Nevertheless, few references focusing on this extraction can be found (Guinda, 2006).

### Activated Carbon

Activated carbon is a solid, which presents a large surface area (high degree of *micro* porosity). This product presents a huge number of industrial, environmental, medical and energy application. It can be produced either by physical or chemical routes (Zabaniotou *et al.*, 2008), but can also be produced by agricultural residues from the oleic sector. In particular, activated carbon is being produced by olive kernels, on an industrial scale. This compound presents maximum BET surface area values (method described by Brunauer, Emmett and Teller to measure the surface area of solids) of up to 1,200 $m^2$ $g^{-1}$ and 3,049 $m^2$ $g^{-1}$ for production on industrial or laboratory scales, respectively (Zabaniotou *et al.*, 2008).

### Ruminant Feeding

Given that olive leaves are fibrous with low digestibility, especially of raw protein and promote very poor rumen fermentation, olive by-products are being evaluated as food for animals. Although dry leaves may be incorporated into the diet, the nutritive value of olive leaves is greater when eaten fresh. Olive leaves are rich in oil, in comparison to diets based on conventional forage, and result in an improvement in milk fat quality mainly for lactating animals. Whereas, extracted olive cake provides cheap energy and fiber for the animal, high-fat olive cake may be used to improve the quality of the fat in animal products. However, more research is needed to assess the potential toxic effect of these potential foods (Molina-Alcaide and Yañez-Ruiz, 2008).

## 2.2. Waste from Three Phase Method

The virgin olive oil produced by the three phase method generates two more outputs, *viz.* solid waste and olive oil mill waste water. In this section, some methods to treat these wastes and several of their potential applications will be proposed here separately.

### 2.2.1. Solid Waste

*Land Application*

This type of application is adequate as an alternative to agricultural wastes. Here, given the nutrient content of the solid waste, it can be applied to enrich the soil for crop production. This application consists of spreading this solid waste on the land and controlling the evolution of the soil and the growth of plants. As this application requires a low energy supply, this process presents high efficiency. However, this spreading of solid waste can be dangerous to animals and some plants and it can also cause acidification of the soil.

*Composting*

The degradation process, from the organic material to humus is called the composting process. This aerobic transformation takes place by enzymatic digestion of the waste which is carried out by microorganisms using the organic matter as substrate. When the waste is received into the factory, it is accumulated in heaps and during the first exothermal stage of manufacture of compost, the temperature inside these heaps can reach up to 80 °C. This heating effect of the solid waste leads to the natural pasteurization of the potential compost. The homogeneous aeration of solid waste guarantees the complete degradation of its *bio*degradable organic matter, which lasts 3 or 4 months. The compost can be used to improve the texture and biological activity for contaminated soils, to control plant diseases because of its high nitrogen and phosphorus content, *bio*fertiliser, with a mixture of compost with commercial substrate finding uses for ornamental plants growth, *bio*filter for toxic metal removal (Arvanitoyannis and Kassaveti, 2007), to control of erosion and landscaping, reforestation, *etc.* This process can reduce the waste mass by up to 50 %. The two main problems of this process are the odor emission and the drainage water that have to be treated. The first problem is satisfactorily solved, by treating the gas with *bio*filters, but the cost of the process is obviously higher.

*Anaerobic Digestion*

The digestion of solid organic wastes is a widely applied technology in the municipal waste treatment field. As the water content of the solid waste is not enough so as to apply this anaerobic fermentation, a pre-treatment consists of hydrating it. The first stage is the acidification of the waste and once the hydrolysis of the organic matter has taken place, methane is produced. Up to 50 % of the organic matter input is transformed in *bio*gas, although the production of low sludge is the main disadvantage.

*Incineration/Combustion*

When a waste cannot be used for any other purpose, burning it is normally the final solution, with the main objective being to recover the most possible heat, to be converted into the maximum possible quantity of electrical energy. Therefore, the incineration of wet waste, where the produced energy is less than the energy consumed, is not recommendable. By incinerating one ton of solid waste, 400,000 kcal can be recovered and contrary to other sectors where the ashes generated are an important environmental problem, in the olive sector, the ashes can be used as a source of mineral for soils. This final application is the only one authorized in specific countries of the European Community. As in the majority of the cases, in addition to environmental impact and risks to human health, directives are more and more restrictive relating to the control of the negative effects on the environment. The main disadvantage is that the amount of energy recovered is low in relation to the investment that needs to be made.

*Pyrolysis*

The thermo chemical conversion of *bio*mass into liquid fuel and chemicals can be carried out by pyrolysis. This process does not involve air emissions. Currently, this liquid fuel presents more interesting advantages related with the storage, transportation and generation of energy than other fuels. As this solid waste presents a sulfur compound concentration of less than 0.01 wt %, it is one of the most suitable wastes to be processed and this olive residue can be considered as a clean energy source. Currently, research works are focused on the design and optimization of reactors where the pyrolysis of olive residues can be carried out with the highest possible yields (Uzun *et al.*, 2007). With this aim, Pütün *et al.* have proposed that using a fixed bed tubular reactor using natural zeolite (clinoptilolite) as the catalyst at 400 °C, a liquid yield higher than 30 % is achieved (Pütün *et al.*, 2009).

*Gasification*

The carbon matter of the solid waste is converted into a synthesis gas composed of carbon monoxide and hydrogen by means of this process. This synthesis gas can be used to generate electricity in an integrated gasification combined cycle or a basic chemical reactant (fuels, tar oils, chemicals or industrial gases). In addition, to synthesis gas, new wastes are generated and this is carried out by means of the gasification of dry solid waste.

### 2.2.2. Olive Oil Mill Waste Water

Olive oil mill waste water (OMW) is mainly formed by water (80 – 95 % by weight). It is characterized by a dark colored liquid caused by lignin polymerized with phenolic compounds, pH values of about 5, and high conductivity, among other factors. Quantitatively, olive oil waste water is formed by 15 % of volatile solid, 2 % of inorganic matter and the remainder formed by organic load. Chemically, olive oil mill waste water is formed by polyphenols, fructose, mannose, glucose, saccharose, sucrose, pentose, tannins, polyalcohols, aromatic compounds, fermentable proteins, organic acids, vitamins, traces of pesticides, a small quantity of emulsified olive oil, *etc.*, see Table 3 (Bazoti *et al.*, 2006; Kapellakis *et al.*, 2008). Olive oil mill waste water chemical composition not only depends on the manufacturing process, but also depends on the variety of olive, stage of maturity, storage time and climate, among others factors. The maximum and minimum ranges of the polluting load, organic constituents and average values of inorganic constituents of olive oil waste water are shown in Table 3. The range can be explained by the aforementioned characteristics and because the olive oil waste water chemical composition changes depending on fermentation reactions during the storage time (Borja *et al.*, 2006).

Given that olive oil mill waste water comes from olives, the *bio*degradability of this waste can be assumed. But because the phenolic compounds degradation rate is much slower than for other substances, olive oil mill waste water is not easily *bio*degradable. On the contrary, for instance, although tannins are highly toxic, these are *bio*degradable. Therefore, due to their high toxicity and their *bio*degradability characteristics, phenolic compounds are the main cause of OMW environmental impact (Kapellakis *et al.*, 2008). During the last few years, the toxicity of phenolic compounds such as tyrosol, hydroxytyrosol, catechol,

protocatechuic acid, caffeic acid, *etc.* have been studied in different seeds
(*Cucumis sativus, Lepidium sativum, Sorghum bicolour, etc.*) and different
organisms (*Daphnia magna, Thamnocephaus platyurus, Brachionus calyciflorus,
Pseudokirchneriella subcapitata, etc.*) (Fiorentino *et al.*, 2003; Isidori *et al.*,
2005). As can be seen in Table 4, the two most toxic phenolic compounds are
cathecol and hydroxytyrosol. In the light of these results, in the case of accidental
introduction of large quantities of OMW into urban sewage treatment plants,
severe phytotoxic effects may occur in the germination and seedling development
(Greco *et al.*, 2006). As Mekki *et al.* stated, the elimination of monomeric
phenolic compounds such as tyrosol and hydroxytyrosol has led to a significant
decrease in toxicity (Mekki *et al.*, 2008), and many treatments of OMW focus on
the removal of these phenolic compounds. Taking into account that the isolate
phenols present beneficial properties for the human circulatory system and some
types of cancer (Visioli and Galli 1998; Salami *et al.*, 1995), proposing reliable
methods to reduce the phenolic concentration is important and the study of OMW
management is essential (*vide infra*).

**Table 3. Olive oil mill wastewater produced by pressure and three phase
centrifugation methods characteristics and their chemical compositions
(Kapellakis *et al.*, 2008).**

|  | Processing system | |
|---|---|---|
|  | Pressure | Three phase centrifugation |
| **Polluting load** | | |
| Chemical oxygen demand (COD) (g·L$^{-1}$) | 120-130 | 60-180 |
| Biological oxygen demand (BOD$_5$) (g·L$^{-1}$) | 90-100 | 20-55 |
| Suspended Solids (%) | 0.1 | 0.9 |
| Total Solids (%) | 12 | 6 |
| **Organic Constituents (%)** | | |
| Sugars | 2.0-8.0 | 0.5-2.6 |
| Nitrogen compounds | 0.5-2.0 | 0.1-0.3 |
| Organic Acids | 0.1-1.5 | 0.2-0.4 |
| Polyalcohols | 1.0-1.5 | 0.3-0.5 |
| Pentoses, Tannins | 1.0-1.5 | 0.2-0.5 |
| Polyphenols | 2.0-2.4 | 0.3-0.8 |
| Lipids | 0.03-1.0 | 0.5-2.3 |
| **Inorganic Constituents (%)** | | |
| Phosphorus | 0.11 | 0.03 |
| Potassium | 0.72 | 0.27 |
| Calcium | 0.07 | 0.02 |
| Magnesium | 0.04 | 0.01 |
| Sodium | 0.09 | 0.03 |
| Chlorine | 0.03 | 0.01 |

**Table 4. How the phenolic compounds (first column) affect microorganisms (first row) is quantified by Median effective concentration expressed as half maximal effective concentration (EC50). These figures signify the molar phenolic compound concentration which produces 50 % of the maximum possible response for the phenolic compounds.**

| | Lepidium sativum[1] | Cucumis sativus[1] | Sorghum bicolour[1] | Sorghum bicolour[1] | Daphnia magna[2] | Thamnocephaus platyurus[2] | Brachionus calyciflorus[2] | Pseudokirchneriella subcapitata[2] |
|---|---|---|---|---|---|---|---|---|
| Catechol | 1.07 | 1.09 | 0.52 | 0.40 | 10 | 8 | 17 | 34 |
| p-hydroxybenzoic acid | 5.36 | 3.55 | 2.56 | 5.36 | 446 | 983 | 225 | 256 |
| Protocatechuic acid | 5.32 | 4.95 | 3.22 | 6.31 | 413 | 589 | 385 | 344 |
| Vanillic acid | 2.65 | 2.04 | 2.05 | 8.79 | 386 | 431 | 1 | 255 |
| Syringic acid | 2.15 | 2.05 | 1.55 | 1.94 | 177 | 97 | 141 | 214 |
| 4-hydroxyphenylacetic acid | 0.86 | 2.68 | 1.04 | 4.13 | 391 | 689 | 273 | 486 |
| 3,4-dihydroxyphenylacetic acid, | 1.10 | 4.40 | 1.97 | 3.87 | 331 | 390 | 136 | 33 |
| Homovanillic acid | 5.23 | 6.62 | 3.30 | 5.23 | 268 | 299 | 407 | 440 |
| Tyrosol | 5.37 | 5.95 | 6.33 | 5.37 | 861 | 296 | 47 | 210 |
| Hydroxytyrosol | 1.02 | 1.55 | 0.47 | 0.82 | 11 | 4 | 9 | 120 |
| 3,4-dihydroxyphenylethylene glycol | 2.22 | 2.30 | 1.08 | 4.14 | 208 | 65 | 144 | 137 |
| p-coumaric acid | 2.03 | 2.11 | 1.15 | 4.05 | 290 | 591 | 108 | 225 |
| Caffeic acid | 9.35 | 11.59 | 1.94 | 2.29 | 326 | 626 | 359 | 120 |
| Ferulic acid | 1.88 | 1.64 | 1.22 | 3.65 | 249 | 300 | 247 | 413 |
| Sinapic acid | 5.39 | 3.66 | 2.68 | 5.51 | 208 | 628 | 398 | 254 |

[1] in mmol·L$^{-1}$, (Isidori et al., 2005)

[2] in μmol·L$^{-1}$, (Fiorentino et al., 2003)

*2.2.2.1. Detoxification Processes*

From 1953, when Professor Fiestas Ros de Ursinos published his first work related with OMW treatment methods, to the present day (Fiestas, 1953), more than 1,000 studies have been published. The detoxification processes can be classified in three wide groups depending on their nature: (*i*) biological; (*ii*) physic-chemical and (*iii*) natural treatments. From the point of view of the reduction of phenolic compounds, all these methods have been briefly revised here.

2.2.2.1.1. Biological Treatments

These methods, mainly based on the aerobic and anaerobic digestion of organic matter of OMW which are used to reduce the phenolic compounds have been shown here.

*Using Aerobic Microorganisms.*

Aerobic bacteria have been tested primarily as a method for the removal of monoaromatic or simple phenolics from olive oil mill waste water. These bacteria appear to be very effective against some phenolic compounds and relatively ineffective against others. For example, *Bacillus pumilus 123* is able to completely degrade protocatechuic and caffeic acids, but has much less affect on tyrosol. In general, using this bacterium, a 50 % reduction in the phenolic content of OMW was found (Ramos-Cormenzana *et al.*, 1996; McNamara *et al.*, 2008).

On the other hand, aerobic bacteria consortia from activated sludge (Borja *et al.*, 1995a; Benitez *et al.*, 1997), commercial communities (Ranalli, 1992), soil, or wastewater (Zouari and Ellouz 1996) have been used to *bio*-remediate olive oil mill waste water. In this *bio*-remediation process, the concentration of phytotoxic compounds and chemical oxygen demand decreases by up to 80 %, and the simple phenolic compounds are completely removed (McNamara *et al.*, 2008).

To reduce the phenolic content of OMW, fungal species such as *Geotrichum candidum, Lactobacillus plantarum, Phanerochaete chrysisporium, Panus tigrinus, etc.* have been used (Kapellakis *et al.*, 2008). Fungal remediation removes simple phenolic compounds and reduces the chemical oxygen demand. Fadil *et al.* have found that using *Geotrichum sp., Aspergillus sp.* and *Candida tropicalis* enables the removal of polyphenols in percentages of 46.6, 44.3 and 51.7 %, respectively (Fadil *et al.*, 2003). Whereas, Tomati *et al.* reported the total removal of phenolic compounds in OMW by means of *Pleorotus ostreatus* (Tomati *et al.*, 1991). Other research groups have found that the percentage of phenolic compound can be reduced by over 90 % using *Pleorotus eyingii, Pleorotus floridae* and *Pleorotus sajor-caju* (Sanjust *et al.*, 1991).

*Using Anaerobic Microorganisms*

By treating OMW using anaerobic methods, two important advantages over the aerobic processes have been found. Firstly, the generation of methane which can be used as an energy source in other processes (Dalis *et al.*, 1996; Erguder *et al.*, 2000), and secondly, this process produces much less waste sludge than aerobic processes. Besides, more than 75 % of toxic phenols and fatty acids are also removed (Dalis *et al.*, 1996; McNamara *et al.*, 2008). An important disadvantage of anaerobic digestion is based on the presence of toxic compounds in the methanogens process of the OMW. In order to overcome this problem, these compounds can be removed by aerobic *pre*-treatment (McNamara *et al.*, 2008).

*Using a Combination of Aerobic and Anaerobic Microorganisms*

Borja *et al.* have demonstrated that aerobic *pre*-treatments increase the production of methane and reduce by up to 23 % the phenolic compounds (Borja *et al.*, 1995b; Borja *et al.*, 1995c). This research group compared anaerobic digestion of OMW *pre*-treated by two different fungi and a bacterium: *Geotrichum candidum*, *Aspergillus terreus* and *Azotobacter chroococcum*. These organisms partially removed the phenolic concentration and therefore the toxicity of OMW is reduced by 59, 87 and 79 %, respectively (Borja *et al.*, 1998). Using *Candida tropicalis* to aerobically *pre*-treat olive oil mill waste water prior to anaerobic digestion resulted in a reduction of 54 % of the phenolic content. *Pre*-treating OMW with *Geotrichum candidum* reduced the phenolic and volatile fatty acid content and increased substrate uptake during anaerobic digestion (Martin *et al.*, 1993).

### 2.2.2.1.2. Physicochemical Treatments

In this group, methods such as distillation, evaporation, combustion or incineration, and flocculation-clarification methods of OMW can be found. These methods can be used mainly to concentrate, or as a final treatment of OMW. Techniques such as adsorption or ion exchange among others are methods that could be used to eliminate phenols and polyphenols but these are usually applied in combination with others (Adhoum and Monser, 2004; Kapellakis *et al.*, 2008). Other methods related to the decrease of the phenolic content of OMW using oxidative processes have been summarized here.

**Fenton reaction** is based on chemical oxidation and coagulation of organic compounds present in OMW by means of hydrogen peroxide and ferrous sulphate. Apart from the chemical oxygen demand reduction, by Fenton reaction, the phenolic content is notably reduced (Azabou *et al.*, 2007). Nevertheless, the total removal of phenolic compounds could be reached using a combination of this reaction and the coagulation process (Vlyssides *et al.*, 2003; Mantzavinos and Kalogerakis 2005).

**Electrochemical oxidation processes**. Given the high conductivity of OMW, their pollutants can be destroyed by means of oxidation processes. The fraction based on aromatic compounds was nearly completely eliminated by Panizza *et al.* (Panizza and Cerisola 2006).

**Advanced oxidation processes**. In this group, oxidation processes based on ozonation are included. Using this method in combination with photocatalysis, UV irradiation and their combinations the total removal of phenol compounds can be achieved (Benitez *et al.*, 1999; Mantzavinos and Kalogerakis 2005).

*2.2.2.1.3.* Natural Treatments

*Land Application*
     Using OMW on soils has two opposite effects. On one hand, considering the fertilizing properties of the waste (potassium, phosphorus, and nitrogen content and that OMW contains neither pathogenic nor heavy metals), spreading OMW on soils enhances fertility. But when the OMW is put onto soil, their salinity and phenolic composition increases and it negatively affects crop production (Sierra *et al.*, 2001; Sierra *et al.*, 2007), although Marsilio *et al.* argued that spreading OMW on cultivated soils does not generate problems (Marsilio *et al.*, 1990). Moreover, some groups stated that OMW spread on soil has an enriching effect (Levi-Minzi *et al.*, 1992; Rouina *et al.*, 1999). Because of this, before applying this treatment, the OMW characteristics, soil properties and the crops involved should be studied in detail, and then, the advantages and disadvantages of natural or conventional treatment should be evaluated (Kapellakis *et al.*, 2008).

*2.2.2.2. Treatments of Olive Oil Mill Waste Water*
     Depending on the region, this waste is called by different names such as aqua *reflue* in Italy, *katsigaros* in Greece, *zebra* in Arab countries, *alpechin* in Spain, *etc.* (Kapellakis *et al.*, 2008). Due to the fact that olive oil mill waste water presents a high organic load (*vide supra*), bactericide substances (polyphenol, oleuropein) and that the main waste load is produced in a short period of time, it should be treated as soon as possible, the treatment of olive oil mill waste waters is very difficult and complex. Currently, there are no universally correct treatments because all techniques in use present various drawbacks.

Separation Treatments

Known techniques (singly or in combination) such as precipitation, filtration, flocculation and coagulation, *etc.* can be found. Usually, after the precipitation process, where dissolved compounds can be precipitated by the addition of chemicals, the filtration process is carried out to clarify and enhance the effectiveness of disinfections of the waste water treated. The separation of solids from olive oil mill waste water is done by means of a porous medium, screen or filter cloth, which retains the solids and allows the liquid to pass through. Usually, the filters are of sand, gravel, and charcoal or activated carbons that help remove even smaller particles. The particles retained and their respective filter types are shown in figure 2. This process is suitable to treat non *bio*degradable substances, and this technique is normally the basis of pre-treatments or post-treatment, i.e. before or after of the principal treatment.

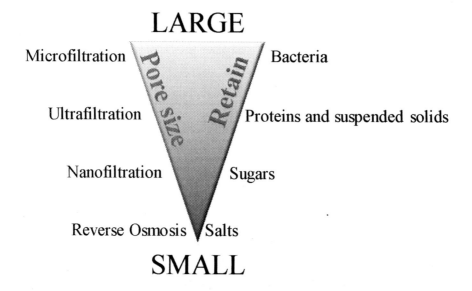

Figure 2. Types of filtration and their respective separable compounds.

Activated Sludge Treatment

This type of treatment was proposed at the beginning of the twentieth century, and currently is being applied in the treatment of industrial, agricultural or municipal waste water. In this case, the process takes place in an aeration tank (closed reactor) where a power agitation between olive oil mill waste water, air and microbial *bio*mass (bacteria) takes place. This *bio*mass is responsible for the degradation process of the waste water and during this process carbon dioxide and ammonia are generated. Part of the output product from the aeration tank is recycled into the input of the tank and the other part must be treated. Schematic diagram of the activated sludge treatment applied to olive oil mill waste water is shown in figure 3. This treatment is used to decrease the biological and chemical oxygen demand. As this process is suitable for waste water with BOD < 3 g L$^{-1}$ and olive oil mill waste water presents BOD > 20 g L$^{-1}$, this process is usually used as a secondary method. Currently, *bio*reactors based on membranes where the microfiltration or even *ultra*filtration processes take place have been tested. With this technology, the equipment is made more compact avoiding the need for a sedimentation tank.

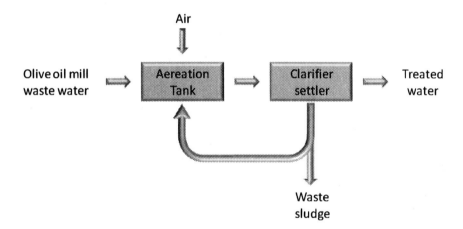

Figure 3. Schematic diagram of the activated sludge treatment.

Incineration

Depending on the percentage of organic matter in olive oil mill waste water, its complete oxidation with air at high temperatures accompanied by a complete evaporation of the water is an adequate alternative to treat this waste. As in the incineration of other raw material, the main problem is ash and the exhaust gas generated. Here, the main disadvantage in incinerating olive oil mill waste water, is its high energy consumption.

Distillation or Evaporation

The separation of groups of compounds from olive oil mill waste water can be carried out by the distillation or the evaporation process. These require a heat flow to vaporize and remove one or more compounds from the mixture. When one of the separated chemicals is thermo sensible, these processes can be carried out under vacuum conditions, although these techniques are expensive, and these are more suitable when the compound separated presents a high economic value (Brenes *et al.*, 1999).

## 2.3. Waste generated from the two phase method

Olive oil extraction by the two-phase method generates 20 kg of olive oil and 80 kg of olive pomace for every 100 kg of fresh olives. The latter is composed approximately of 55 % water (all the vegetation water and sometimes water added during the manufacturing process), 22 % pulp, 20 % olive pits and 3 % fat content. Some characteristics of olive pomace are shown in Table 1 (Caputo *et al.*, 2003; Torrecilla *et al.*, 2006); especially interesting is the low sulfur content (in view of the further use of the de-oiled olive pomace as clean fuel), the content of potassium and phosphor (in view of the use of combustion ash in fertilization) and the high content of sugars, which leads to problems inside rotary dryers (Torrecilla, 2001; Torrecilla, 2001a; Caputo *et al.*, 2003; Torrecilla *et al.*, 2006; Torrecilla *et al.*, 2005).

### 2.3.1. Olive Oil Pomace

Given the great variety of compounds present in this waste, there are many techniques to remove those environmental harmful compounds. Some of them are in the research stage in laboratories and other are being industrially applied. From an industrial point of view, the olive pomace is used in different industrial fields such as energy, chemical, environmental, *etc.* (Torrecilla, 2001a). In particular, in the industrial sector, the widest applications can be classified into different groups (Arvanitoyannis *et al.*, 2007a): **Bio-remediation** where the metabolic potential of microorganisms to clean the olive pomace is used (Arvanitoyannis and Kassaveti, 2007). These processes can be carried out under aerobic or anaerobic conditions. These types of processes can be applied *in-situ* or can be transported to another location before treatment (*ex-situ*). Another broad group is the **thermal process** such as incineration, drying, pyrolysis, gasification or evaporation, where heat is applied to stabilize the waste and depending on the process, energy can be recovered or not. A different type of application is based on the **membrane processes.** Using this technology the waste can be concentrated or diluted depending on its molecular size, figure 2. This technology has been established and applied in many industrial fields for the treatment of different types of waste. Among other advantages, low energy cost, reliability and a reduction of capital outlay can be highlighted. Another group of applications using electricity is composed of **electrolysis or ozonation**. The result of these is less toxic waste with an economic value. Another group composed of **digestion, coagulation/ flocculation/ precipitation processes** present the same characteristics, modifying the molecular size by breaking or agglomerating molecules. Generally, these types of processes are highly effective. And finally **distillation processes** are also used to separate hazardous chemicals from the initial waste (Arvanitoyannis *et al.*, 2007a). Some of them are described in more detail below.

Figure 4. Rotary dryer used to dry a large part of the olive pomace produced in the south of Córdoba, Spain during the 1996/97 season (Oleícola El Tejar Nuestra Señora de Araceli Sdad. Coop. Ltda.).

*Drying of Olive Pomace*

Olive pomace from the two-phase method of olive oil extraction (two-phase olive pomace) must be dried from about 65 % [wet basis (wb-%)] to about 8 %, in order to make possible the extraction of the remaining oil (about 3 wb-%) or be applied to other objectives. This operation is usually carried out in a rotary drier. The rotary dryer used to dry all olive pomace generated in the south of Córdoba in Spain is shown in figure 4 (Oleícola El Tejar Nuestra Señora de Araceli Sdad. Coop. Ltda.). Krokida *et al.* proposed a design for an appropriate industrial rotary dryer and discussed economic feasibility (Krokida *et al.*, 2002). Also, this operation has been carried out in a fluidized bed, fluidized/moving bed drier on a small industrial scale (Torrecilla, 2004). By the fluidized drier, the dried material is homogeneous and contains a negligible amount of polycyclic aromatic hydrocarbons (Torrecilla et al., 2006). In the fluidized/moving bed drier, the moving bed acts as a pre-dryer of the wet solid and as a filter of the output gas, retaining more than 99.9 % of the fine particles. The mean power consumption of the fluidized/moving bed drier plant is 1 kWh $kg_{water}^{-1}$; which means a significant reduction of power cost with respect to the rotary dryers, which require more than 1.4 kWh $kg_{water}^{-1}$ (Torrecilla, 2004).

*De-Oiling of Olive Pomace*

Once the virgin olive oil has been separated from the olive paste, a waste with an oil content of up to 3 % is generated. The de-oiling process consists of physically extracting olive oil remains in the waste by a two phase decanter (second centrifugation operation). Obviously, this second olive oil presents a poorer quality than that produced by the first centrifugation. In this field, some research has been carried out to optimize the operation conditions e.g. with/without pit, malaxing time, enzymes, water addition, milling or combinations of them (Koprivnjak et al., 2009; López-Piñeiro et al., 2008; Aragón and Palancar, 2001).

*Gasification of Olive Pomace*

Carbonaceous substances such as coal, petroleum or wastes such as *bio*mass, olive pomace or deoiled orujo (waste generated in the manufacture of refined olive-pomace oil, so called orujillo in Spanish) can be converted into carbon monoxide and hydrogen. This mixture is commonly called synthesis gas. This gas can be used for heat production and for the generation of mechanical and electrical power. Given the moisture and rheological characteristics of the olive pomace, the gasification of this waste is difficult. However, successful results of experimental tests have shown that the gasification of de-oiled orujo (carbon content of 45 % in wet basis) is feasible. The flue gas obtained has a typical composition and, by further testing its use in internal combustion engines to produce electricity could prove feasible (Aragón and Palancar, 2001). Currently, designs of specific reactors have been developed in order to take as much advantage as possible of the energy and the synthesis gas produced (Palancar et al., 2009).

# 3. WASTE GENERATE DURING THE TABLE OLIVE MANUFACTURE

As was presented in Chapter 3, the manufacture of table olives depends on many geographical, cultural factors and others related with the olive variety. Although the compositions of the waste can vary appreciably, the most typical characteristics of waste water streams from the manufacture of table olives are shown in table 5. Although the productive process of table olive is being continuously improved, the manufacturing process consumes 1,200 m$^3$ t$_{olives}^{-1}$ of fresh water. Given that the 2008/09 season a production of more than $2 \cdot 10^6$ tones of table olives is expected an enormous quantity of contaminated water will be generated in the world, and therefore effective methods are required.

**Table 5. Main characteristics of waste water from table olives.**

| Characteristics | Lye and wash water | Brine |
|---|---|---|
| pH | 9-13 | 4 |
| Sodium hydroxide (g L$^{-1}$) | 1.1-1.5 | - |
| Sodium chloride (g L$^{-1}$) | - | 6-10 |
| Free acidity (g$_{lactic\ acid}$ L$^{-1}$) | - | 6-15 |
| Polyphenols (g$_{tannic\ acid}$ L$^{-1}$) | 4.1-6.3 | 5-7 |
| Chemical oxygen demand (g O$_2$ L$^{-1}$) | 23-28 | 10-20 |
| Biological oxygen demand (g O$_2$ L$^{-1}$) | 15-25 | 9-15 |
| Volatile organic solids (g L$^{-1}$) | 30-40 | 10-20 |

In order to propose the most appropriate method to treat this waste, methods which use less energy, less water and produce lower volumes of waste-water are required. Methods based on natural fermentation in brine or dry salt are more favorable than others involving heat, which increase general costs and possibly labor costs. Taking into account that to produce edible olives, aqueous solutions are involved in nearly all stages, *viz.* processing with water, brine and lye (running water, sodium chloride and sodium hydroxide aqueous solutions, see Chapter 3). The techniques to treat the wastes generated during the manufacture of table olives can be classified into three groups *viz.* physical processes, physicochemical process and their combination.

## 3.1. Physical Processes

### Filtration

To separate organic compounds from the brine, *micro*, *ultra* and even *nano*filtration can be used, figure 2. Another method to reduce the waste amount generated is the recycling of the waste water, and employing as many cycles as possible. Immediately after the treatment, this regeneration can be carried out by *ultra*fitration, although meets with two important problems *viz.* darkening of the solutions and the growth of yeast and mold where the surface is in contact with air. Working at 30-40 ºC is recommendable to solve this problem. In California, a company which manufactures table olives has included a filtration stage in the waste treatment line. In this factory, every day 264 m$^3$ of waste are treated. Previous to recycling this waste, the fluid is *ultra*filtered, and then, reverse osmosis is subsequently applied. The concentrate retained is dehydrated by evaporation. This alternative was selected, mainly because of its cost and energy consumption (Niaounakis and Halvadakis, 2006).

## 3.2. Physicochemical Processes

### Flocculation/Adsorption

These processes are used to decontaminate table olive processing wastewater, reducing up to 40 % of the initial pollution. As with stages prior to these processes, to favor the decantation and reduce the size of the plant, the adjustment of adequate operational conditions (pH, temperature, *etc.*) is required. In patent number US3975270 the purification process of this waste is described, which consists of four stages: firstly, lime, charcoal and calcium carbonate are added sequentially; then, the blend formed is mixed for one hour; followed by the separation of the usable phase; and finally the aqueous phase is recovered by filtering (Niaounakis and Halvadakis, 2006). However, reconditioned brines of lower salt concentration were reused to store olives with no detectable effect on the product quality. These brines were produced by the treatment of this waste with activated carbon.

### Chemical Oxidation Processes

To reduce the chemical oxygen demand and the polyphenols content, and also to increase the *bio*degradability of the waste, chemical oxidation by means of ozone or advanced oxidation technologies can be used. An ozone dose of 45 mg L$^{-1}$ during 35 minutes is adequate to achieve a decrease of pH and phenolic content and increases the *bio*degradability of the waste. The aromatic content and color almost completely disappear using doses less than 0.5 g of ozone. Segovia *et al.* have studied the effect of ozone on the fermentation brines from Spanish green table olive processing, and the recycling capability of the fermentation brine; the physicochemical and organoleptic characteristics of the table olives. The ozone required for the complete elimination of polyphenolic compounds depends on the pH, i.e., at pH 4 and 10 units, 15 and 7 mg $L^{-1}$ of ozone are required, respectively. The treated brines are then filtered through a 0.45 $\mu$m pore size. Finally, the table olives are ready for sale (Segovia-Bravo *et al.*, 2008).

The oxidation of table olive processing waste water by means of advanced oxidation technology can be carried out by a combination of UV radiation and hydrogen peroxide, using Fenton's reaction or by photo-Fenton. By means of Fenton's reaction, the treatment is done by a powerful oxidizing agent which is capable of degrading a wide number of organic and even inorganic substances. In addition, the chemical oxygen demand can be reduced by up to 90 % and moderate reduction of total carbon (up to 60 %) by the combination of ozone (doses of 3-4 g), 254 nm UV radiation or hydrogen peroxide (initial concentration of $10^{-3}$ M) can be achieved. One of the most effective processes was the combination of ozone, UV radiation and hydrogen peroxide (Niaounakis and Halvadakis, 2006).

Another combination of methods with a successful result was described by Kotsou *et al.* in which wastewater from the debittering process of green table olives is treated aerobically using *Aspergillus niger* during two days. The main results consist of a reduction of COD by 70 % and the total and simple phenolic compounds were reduced by 41 and 85 %, respectively. Then, to oxidize the recalcitrant organic compounds and/or metabolites of those that could not oxidize biologically Fenton's reagent was used as a secondary chemical treatment (Kotsou *et al.*, 2004).

## 3.3. Combined Processes

Research groups stated that the combination of various processes can degrade the toxic and non-*bio*degradable organic compounds in table olives processing waste water more efficiently than when applied individually. The treatment by activated carbon in combination with *ultra*filtration (polysulfone membrane) or flocculation has been satisfactorily tested for the fermentation brines of Spanish style green table olives (Romero *et al.*, 2001; Niaounakis and Halvadakis, 2006). The combination of fermentation and evaporation under vacuum condition processes for the treatment by washing of Spanish style green olives has been tested. The water used for washing was fermented in a 500 L tank and when the process was tested in a pilot plant the COD showed a reduction of up to 70 % (Brenes *et al.*, 2004).

## 4. CONCLUSIONS

In this Chapter, different techniques to treat the residues generated during the manufacture of virgin olive oil and table olive have been presented. As this oil is produced mainly by means of two phase and three phase methods, the olive pomace and olive oil mill waste water and solid residues have been studied. Relating to table olives, techniques of treatment based on physical property, physicochemical characteristics and the combinations of both types of processes have been shown.

To reduce the concentration of phenolic compounds, therefore reducing its environmental impact, and for appropriate management of olive oil mill waste water, the best method to be applied still requires further detailed study. To improve extra virgin olive oil and table olive quality and reduce the environmental impact of wastes generated during their production processes, a detailed study of equipment and appropriate management of their wastes are required. Because the most suitable method depends on many factors (cultivar, water amount available, region, climatology, economy, *etc.*), by taking them all into consideration, the most adequate method or combination of them can be selected.

## REFERENCES

Adhoum, N. & Monser, L. (2004). Decolourization and removal of phenolic compounds from olive mill wastewater by electrocoagulation. *Chemical Engineering and Processing, 43*, 1281-1287.

Aragón, J. M., & Palancar, M. C. IMPROLIVE 2000, present and future of alpeorujo. Madrid: Ed. Complutense S. A.; 2001.

Arvanitoyannis, I. S. & Kassaveti, A. (2007). Current and potential uses of composted olive oil waste. *International Journal of Food Science and Technology*, 42, 281–295.

Arvanitoyannis, I. S., Kassaveti, A. & Stefanatos, S. (2007). Olive Oil Waste Treatment: A Comparative and Critical Presentation of Methods, Advantages & Disadvantages. *Reviews in Food Science and Nutrition, 47*, 187–229.

Azabou, S., Najjar, W., Gargoubi, A., Ghorbel, A. & Sayadi, S. (2007). Catalytic wet peroxide photo-oxidation of phenolic olive oil mill wastewater contaminants - Part II. Degradation and detoxification of low-molecular mass phenolic compounds in model and real effluent. *Applied Catalysis B: Environmental, 77*, 166-174.

Baldioli, M., Servili, M., Perretti, G. & Montedoro, G. F. (1996). Antioxidant Activity of Tocopherols and Phenolic Compounds of Virgin Olive Oil. *Journal of the American Oil Chemists' Society, 73*, 1589-1593.

Bazoti, F. N., Gikas, E., Skaltsounis, A. L. & Tsarbopoulos, A. (2006). Development of a liquid chromatography–electrospray ionization tandem mass spectrometry (LC–ESI MS/MS) method for the quantification of bioactive substances present in olive oil mill wastewaters. *Analytica Chimica Acta, 573–574*, 258–266.

Benitez, F. J., Beltran-Heredia, J., Torregrosa, J. & Acero, J. L. (1999). Treatment of olive mill wastewaters by ozonation, aerobic degradation and the combination of both treatments. *Journal of Chemical Technology and Biotechnology, 74*, 639–46.

Benitez, J., Beltran-Heredia, J., Torregrosa, J., Acero, J. L. & Cercas, V. (1997). Aerobic degradation of olive mill wastewaters. *Applied Microbiology and Biotechnology, 47*, 185–188.

Borja, R., Alba, J. & Banks, C. J. (1995a). Activated sludge treatment of wash waters derived from the purification of virgin olive oil in a new manufacturing process. *Journal of Chemical Technology and Biotechnology, 64*, 25–30.

Borja, R., Alba, J., Garrido, S. E., Martinez, L., García, M. P., Monteoliva, M., & Ramos-Cormenzana, A. (1995b). Effect of aerobic pretreatment with Aspergillus terreus on the anaerobic digestion of olive-mill wastewater. *Biotechnology and Applied Biochemistry, 22*, 233–246.

Borja, R., Alba, J., Garrido, S. E., Martinez, L., García, M. P., Incerti, C. & Ramos-Cormenzana, A. (1995c). Comparative study of anaerobic digestion of olive mill wastewater (OMW) and OMW previously fermented with aspergillus terreus. *Bioprocess and Biosystems Engineering, 13*, 317–322.

Borja, R., Alba, J., Mancha, A., Martin, A., Alonso, V. & Sanchez, E. (1998). Comparative effect of different aerobic pretreatments on the kinetics and macroenergetic parameters of anaerobic digestion of olive mill wastewater in continuous mode. *Bioprocess and Biosystems Engineering, 18*, 127–134.

Borja, R., Raposo, F. & Rincón, B. (2006). Treatment technologies of liquid and solid wastes from two-phase olive oil mills. *Grasas y Aceites, 57*, 32-46.

Brenes, M, Romero, C. & de Castro, A. (2004). Combined fermentation and evaporation processes for treatment of washwaters from Spanish-style green olive processing. *Journal of Chemical Technology and Biotechnology, 79*, 253-259.

Brenes, M., García, A., García, P., Rios, J. J. & Garrido, A. (1999). Phenolic Compounds in Spanish Olive Oils. *Journal of Agricultural and Food Chemistry, 47*, 3535-3540.

Caputo, C., Scacchia, F. & Pelagagge, P. M. (2003). Disposal of byproducts in olive oil industry: Waste-to-energy solutions. *Applied Thermal Engineering, 23*, 197-214.

Dalis, D., Anagnostidis, K., Lopez, A., Letsiou, I. & Hartmann, L. (1996). Anaerobic digestion of total raw olive-oil wastewater in a two-stage pilot-plant (up-flow and fixed-bed *bio*reactors). *Bioresource Technology, 57*, 237–243.

Dalis, D., Anagnostidis, K., Lopez, A., Letsiou, I., & Hartmann, L. (1996). Anaerobic digestion of total raw olive-oil wastewater in a two-stage pilot-plant (up-flow and fixed-bed *bio*reactors). *Bioresource Technology, 57*, 237–243.

Erguder, T. H., Guven E, & Demirer, G. N. (2000). Anaerobic treatment of olive mill wastes in batch reactors. *Process Biochemistry, 36*, 243–248.

Fadil, K., Chahlaouia, A., Ouahbib, A., Zaida, A., & Borja, R. (2003). Aerobic biodegradation and detoxification of wastewaters from the olive oil industry. *International Biodeterioration and Biodegradation, 51*, 37-41.

Fiestas, J. A. (1953). Estudio del alpechín para su aprovechamiento industrial I. concentración de los azúcares y demás substancias que lleva en emulsión y disolución por tratamiento con óxido de calcio. *Grasas y Aceites, 4*, 63–67.

Fiorentino, A., Gentili, A., Isidori, M., Monaco, P., Nardelli, A., Parrella, A. & Temussi, F. (2003). Environmental effects caused by olive mill wastewaters: toxicity comparison of low-molecular-weight phenol components. *Journal of Agricultural and Food Chemistry, 51*, 1005-1009.

Greco, G., Colarieti, M. L., Toscano, G., Iamarino, G., Rao, M. A. & Gianfreda, L. (2006). Mitigation of olive mill wastewater toxicity. *Journal of Agricultural and Food Chemistry, 54*, 6776-6782.

Guinda, A. (2006). Use of solid residue from the olive industry. *Grasas y Aceites, 57*, 107-115.

International Olive Council. 2008. (Last visited August 2009) URL http://www.internationaloliveoil.org.

Isidori, M., Lacorgna, M., Nardelli, A. & Parrella, A. (2005). Model study on the effect of 15 phenolic olive mill wastewater constituents on seed germination and vibrio fischeri metabolism. *Journal of Agricultural and Food Chemistry, 53*, 8414-8417.

Kapellakis, I. E., Tsagarakis, K. P. & Crowther, J. C. (2008). Olive oil history, production and by-product management. *Reviews in Environmental Science and Biotechnology, 7*, 1–26.

Koprivnjak, O., Majetić, V., Staver, M.M., Lovrić, A. & Blagović, B. (2009). Effect of phospholipids on extraction of hydrophilic phenols from virgin olive oils. *Food Chemistry*, doi: 10.1016/j.foodchem.2009.07.016.

Kotsou, M., Kyriacou, A., Lasaridi, K. & Pilidis, G. (2004). Integrated aerobic biological treatment and chemical oxidation with Fenton's reagent for the processing of green table olive wastewater. *Process Biochemistry, 39*, 1653–1660.

Krokida, M. K., Maroulis, Z. B. & Kremalis, C. (2002). Process design of rotary dryers for olive cake. *Drying Technology, 20*, 771–787.

Le Tutour, B. & Guedon, D. (1992). Antioxidative activities of Olea europaea leaves and related phenolic compounds. *Phytochemistry, 31*, 1173-1178.

Levi-Minzi, R., Saviozzi, A., Riffaldi, R. & Falzo, L. (1992). Land application of vegetable water: effects on soil properties. *Olivae, 40*, 20–25.

López-Piñeiro, A., Fernández, J., Albarrán, A., Rato Nunes, J. M. & Barreto, C. (2008). Effects of De-oiled Two-Phase Olive Mill Waste on Mediterranean Soils and the Wheat Crop. *Soil Science Society of America Journal, 72*, 424-430.

Mantzavinos, D. & Kalogerakis, N. (2005). Treatment of olive mill effluents Part I. Organic matter degradation by chemical and biological processes—an overview. *Environment International, 31*, 289– 295.

Marco, E. D., Savarese, M., Paduano, A. & Sacchi, R. (2007). Characterization and fractionation of phenolic compounds extracted from olive oil mill wastewaters. *Food Chemistry*, 104, 858–867.

Marsilio, V., Di Giovacchino, L., Lombardo, N. & Briccoli-Bati, C. (1990). First observations on the disposal effects of olive mills vegetation waters on cultivated soil. *Acta Horticulturae, 286*, 493–498.

Martin, A., Borja, R. & Chica, A. (1993). Kinetic study of an anaerobic fluidized bed system used for the purification of fermented olive mill wastewater. *Journal of Chemical Technology & Biotechnology, 56*, 155–162.

McNamara, C. J., Anastasiouc, C. C., O'Flahertyd, V. & Mitchell, R. (2008). Bioremediation of olive mill wastewater. *International Biodeterioration and Biodegradation, 61*, 127–134.

# 120 José S. Torrecilla

Mekki, A., Dhouib, A., Feki, F., & Sayadi, S. (2008). Assessment of toxicity of the untreated and treated olive mill wastewaters and soil irrigated by using microbiotests. *Ecotoxicology and Environmental Safety, 69*, 488–495.

Molina-Alcaide, E. & Yañez-Ruiz, D. R. (2008). Potential use of olive by-products in ruminant feeding: A review. *Animal Feed Science and Technology, 147*, 247–264.

Niaounakis, M., & Halvadakis, C. P. Olive Processing Waste Management - Literature Review and Patent Survey. 2nd Edition. Pergamon. Amsterdam. 2006.

Obied, H. K., Allen, M. S., Bedgood, D. R., Prenzler, P. D., Robards, K., & Stockmann, R. (2005). Bioactivity and analysis of biophenols recovered from olive mill waste. *Journal of Agricultural and Food Chemistry, 53*, 823-837.

Palancar, M. C., Serrano, M. & Aragón, J. M. (2009). Testing the technological feasibility of FLUMOV as gasifier. *Powder Technology, 194*, 42–50.

Panizza, M. & Cerisola, G. (2006). Olive mill wastewater treatment by anodic oxidation with parallel plate electrodes. *Water Research, 40*, 1179–1184.

Pizzichini, M., & Russo, C. (2005). Process for recovering the components of olive mill wastewater with membrane technologies. Patent No. WO2005123603.

Pütün, E., Uzun, B. B. & Pütün, A. E. (2009). Rapid Pyrolysis of Olive Residue. 2. Effect of Catalytic Upgrading of Pyrolysis Vapors in a Two-Stage Fixed-Bed Reactor. *Energy & Fuels, 23*, 2248–2258.

Ramos-Cormenzana, A., Juarez-Jimenez, B., & García-Pareja, M. P. (1996). Antimicrobial activity of olive mill waste-waters (alpechín) and biotransformed olive oil mill wastewater. *International Biodeterioration and Biodegradation, 38*, 283–290.

Ranalli, A. (1992). Microbiological treatment of oil mill waste waters. *Grasas y Aceites, 43*, 16–19.

Romero Barranco, C., Brenes Balbuena, M, García García, P. & Garrido Fernández, A. (2001). Management of spent brines of osmotic solutions. *Journal of Engineering, 49*, 237-246.

Rouina, B. B., Taamallah, H. & Ammar, E. (1999). Vegetation water used as a fertilizer on young olive plants. *Acta Horticulturae, 474*, 353–355.

Salami, M., Galli, C., De Angelis, L., & Visioli, F. (1995). Formation of F2-isoprostanes in oxidized low density lipoprotein: inhibitory effect of hydroxytyrosol. *Pharmacological Research, 31*, 275–279.

Sanjust, E., Pompei, R., Rescigno, A., Rinaldi, A., & Ballero, M. (1991). Olive milling wastewater as a medium for growth of four Pleurotus species. *Applied Biochemistry and Biotechnology, 31*, 223–235.

Segovia-Bravo, K. A., García-García, P., Arroyo-López, F. N., López-López, A. & Garrido-Fernández, A. (2008). Ozonation process for the regeneration and recycling of Spanish green table olive fermentation brines. *European Food Research and Technology, 227*, 463–472.

Sierra, J., Martí, E., Garau, M. A. & Cruañas, R. (2007). Effects of the agronomic use of olive oil mill wastewater: Field experiment. *Science of the Total Environment, 378*, 90–94.

Sierra, J., Martí, E., Montserrat, G., Cruañas, R. & Garau, M. A. (2001). Characterisation and evolution of a soil affected by olive mill wastewater disposal. *Science of the Total Environment, 279*, 207–214.

Tomati, U., Galli, E., Di Lena, G., & Buffone, R. (1991). Induction of laccase in Pleurotus ostreatus mycelium grown in olive oil waste waters. *Agrochimica 35*, 275–279.

Torrecilla, J. S. (2001). Aprovechamiento del alpeorujo. *Agricultura, revista agropecuaria, 832*, 734-737.

Torrecilla, J. S. (2001a). Secado del orujo en lecho fluidizado / móvil. *Agricultura, Revista Agropecuaria, 827*, 340-343.

Torrecilla, J. S. Secado del orujo en lecho fluidizado movil. Madrid: Ed. Complutense S.A.; 2004.

Torrecilla, J. S., Aragón, J. M. & Palancar, M. C. (2005). Modeling the drying of a high-moisture solid with an artificial neural network. *Industrial Engineering & Chemistry Research, 44*, 8057–8066.

Torrecilla, J. S., Aragón, J. M. & Palancar, M. C. (2006). Improvement of fluidized bed dryers for drying olive oil mill solid waste (olive pomace). *European Journal of Lipid Science and Technology, 108*, 913–924.

Uzun, B. B., Pütün, A. E. & Pütün, E. (2007). Rapid Pyrolysis of Olive Residue. 1. Effect of Heat and Mass Transfer Limitations on Product Yields and Bio-oil Compositions. *Energy & Fuels, 21*, 1768-1776.

Viesca, R. D. L., Fernández, E. & Salvador, J. (2007). Analysis of the scientific production of olive products. I – Table olives. *Grasas y aceites, 58*, 307-310.

Villanova, L., Villanova, L., Fasiello, G., & Merendino, A. (2006). Process for the recovery of tyrosol and hydroxytyrosol from oil mill wastewaters and catalytic oxidation method in order to convert tyrosol in hydroxytyrosol. Patent No. US 2006/0070953.

Visioli, F., & Galli, C. (1998). Olive oil phenols and their potential effects on human health. *Journal of Agricultural and Food Chemistry, 46*, 4292-4296.

Vlyssides, A., Loukakis, H., Israilides, C., Barampouti, E. M. & Mai, S. Detoxification of olive mill wastewater using a Fenton process. Crete: 2nd European bioremediation Conference (N. Kalogerakis, Ed.); 2003.

Zabaniotou, A., Stavropoulos, G. & Skoulou, V. (2008). Activated carbon from olive kernels in a two-stage process: Industrial improvement. *Bioresource Technology, 99*, 320–326.

Zouari, N. & Ellouz, R. (1996). Microbial consortia for the aerobic degradation of aromatic compounds in olive oil mill effluent. *Journal of Industrial Microbiology and Biotechnology, 16*, 155–162.

# INDEX

## A

abiotic, 14
absorbents, 86
absorption, 13, 73
accidental, 103
acetate, 70
acid, 6, 13, 26, 36, 37, 38, 66, 68, 69, 70, 73, 81, 87, 91, 92, 103, 104, 106, 113
acidification, 100, 101
acidity, 36, 38, 68, 70, 71, 72, 73, 74, 84, 85, 86, 89, 113
activated carbon, 86, 99, 108, 114, 115
additives, 99
adjustment, 114
adsorption, 106
adulteration, 66, 80, 81, 82, 86
aerobic, 37, 100, 105, 106, 110, 117, 119, 122
aerobic bacteria, 105
age, 6, 7, 18, 19, 20
ageing, 19, 20
agent, 30, 115
agricultural, vii, 11, 21, 99, 100, 108
agricultural residue, 99
agriculture, 4
aid, 56
air, 16, 19, 20, 21, 34, 35, 37, 54, 55, 74, 101, 108, 109, 114
air emissions, 101
alcohol, 42, 59
aldehydes, 70
alkali, 35, 36, 37, 86
alkaline, 28, 30
almonds, 15
alternative, 19, 35, 38, 78, 100, 109, 114
aluminum, 78
amino, 13, 18
ammonia, 18, 109
ammonium, 18
anaerobic, 37, 101, 105, 106, 110, 117, 119
aniline, 66
animals, 5, 99, 100
anomalous, 88
anthocyanin, 37
antioxidant, 59, 73, 75, 76, 78, 81, 96, 99
antioxidative, 119
application, 36, 38, 50, 79, 95, 99, 100, 101, 110, 119
aqueous solution, 37, 38, 113
Arab countries, 107
arid, 13, 20, 36
aromatic compounds, 56, 102, 107, 122
aromatic hydrocarbons, 112
ash, 109, 110
asia, 6, 8, 27

aspergillus niger, 115
aspergillus terreus, 106, 117
assessment, 71
assimilation, 11, 21
athletes, 4
atmosphere, 17, 18, 35, 38, 47, 61
authentication, 82
authenticity, 87

**B**

bacteria, 36, 38, 105, 108
bacterium, 36, 105, 106
bayesian methods, 7, 9
beating, 56, 75
beef, 90
benefits, 78, 79
bible, 4, 47
binary blends, 81
bioactive compounds, 99
biochemistry, 11, 18, 21
biodegradability, 102, 114
biodegradable, 100, 102, 108, 115
biodegradation, 118
biogas, 101
biogeography, 10
biological activity, 100
biological processes, 119
biomass, 101, 108, 112
bioreactors, 109, 118
bioremediation, 121
biotic, 14
birds, 11
bleaching, 83, 84, 85, 86, 87, 91
blends, 81, 91, 92
botulinum, 38
bounds, 86
broad spectrum, vii
bubbles, 37, 89
burning, 101
by-products, 99, 120

**C**

caffeic acid, 103, 105
calcium, 114
calcium carbonate, 114
calorimetry, 80
cancer, 103
candida, 105, 106
carbohydrates, 13, 14, 18, 19, 85
carbon, 13, 16, 17, 34, 35, 37, 86, 89, 91, 98,
    99, 102, 108, 112, 114, 115, 122
carbon dioxide, 16, 17, 34, 35, 37, 89, 91, 108
carbon monoxide, 102, 112
carboxylic, 26
carotene, 81
carotenoids, 86, 87
catabolic, 55, 74
catalyst, 102
catechol, 102
cell, 14
cellulose, 15, 16
charcoal, 108, 114
chemical composition, 34, 66, 70, 73, 78, 79,
    84, 102, 103
chemical oxidation, 106, 114, 119
chemical reactions, 52, 86, 87
chemicals, vii, 11, 13, 16, 17, 18, 32, 72, 83,
    84, 85, 86, 87, 89, 101, 102, 108, 109, 111
chemometrics, 81
cherries, 15
chloride, 35, 36, 37, 38, 43, 113
chlorophyll, 16, 17, 76, 78, 86
chromatography, 81, 117
chromosome, 12
circulation, 19, 20, 21, 54
classical, vii, viii, 9, 26, 57
classification, 16, 28, 30, 31, 37, 41, 65, 68,
    79, 81, 87
clean energy, 101
cleaning, 55, 56, 74, 76
climatology, 9, 12, 116
clostridium botulinum, 38

clustering, 7
coagulation, 106, 108, 111
coagulation process, 107
coal, 112
coil, 58
colors, 14, 30
combustion, 106, 110, 112
commercialization, 78, 84
community, 39, 105
competition, 19
complexity, 78
components, 14, 76, 85, 89, 96, 118, 120
composition, 6, 15, 31, 34, 39, 41, 70, 73, 78, 79, 80, 81, 84, 87, 91, 102, 107, 112
composting, 95, 96, 100
compounds, 16, 17, 18, 28, 35, 36, 43, 56, 58, 61, 67, 70, 72, 75, 76, 78, 81, 83, 85, 86, 89, 91, 99, 102, 103, 104, 105, 106, 107, 108, 109, 110, 114, 115, 116, 117, 119, 122
concentration, 14, 17, 18, 25, 34, 35, 36, 37, 41, 59, 65, 66, 70, 71, 72, 75, 76, 84, 85, 87, 88, 89, 90, 99, 101, 103, 104, 105, 106, 114, 115, 116
concordance, 21
conditioning, 92
conductivity, 102, 107
confusion, 30
congress, iv
conjugation, 86
conservation, 38, 76
consumers, 31, 38, 39, 65, 66, 79
consumption, 20, 28, 31, 36, 37, 39, 40, 49, 59, 66, 67, 78, 79, 96, 109, 112, 114
contaminants, 72, 117
contaminated soils, 100
control, 19, 81, 100, 101
conversion, 101
cooking, 90
corn, 66
cosmetics, 5
costs, 20, 54, 57, 96, 113
coupling, 82
crop production, 100, 107

crops, 107
crude oil, 81
crystallization, 77, 88
cultivation, vii, 7, 8, 9, 11, 20, 49, 51, 79, 97
cultural factors, 113
culture, 21
curing, 28, 35, 43
cuticle, 14
cutters, 55
cycles, 18, 51, 114
cytoplasm, 14

**D**

death, 66
decomposition, 78
defects, 25, 31, 32, 34, 41, 54, 68, 70, 71
degradation, 34, 36, 55, 67, 73, 74, 78, 79, 100, 102, 108, 117, 119, 122
degradation process, 36, 79, 100, 108
degradation rate, 73, 102
degrading, 115
delivery, 74
denitrification, 18
density, 48, 57, 120
density values, 57
deodorizing, 84
Department of Agriculture, 65, 67, 68, 79, 82
designers, 61
detection, 81
detoxification, 95, 105, 117, 118, 121
dienes, 86
diet, 3, 4, 5, 7, 27, 99
differentiation, 14
digestibility, 99
digestion, 100, 101, 105, 106, 111, 117, 118
diploid, 11
directives, 101
discharges, 18
discs, 60
diseases, 16, 19, 100
distillation, 86, 87, 106, 109, 111

DNA, 21
domestication, 7, 10
draft, 68
drainage, 101
drought, 14
drying, 28, 63, 110, 121
duration, 75

## E

earth, 85, 86
eating, 66
ecological, 97
ecology, 21
economics, 4, 42
effluent, 97, 117, 119, 122
elaboration, 25, 31
electrical power, 47, 112
electricity, 9, 102, 111, 112
electrodes, 120
electrolysis, 111
emission, 101
energy, 9, 16, 57, 59, 95, 99, 100, 101, 106,
    109, 110, 112, 113, 114, 118
energy consumption, 59, 109, 114
energy supply, 100
engines, 112
enlargement, 13
environment, 16, 95, 101
environmental conditions, 12, 72
environmental impact, vii, 95, 96, 97, 101,
    102, 116
enzymatic, 18, 55, 71, 74, 75, 78, 90, 100
enzymes, 17, 56, 71, 112
eritrea, 27
erosion, 101
ESI, 117
ester, 26
esterification, 90
estimating, 78
ethyl acetate, 70
ethylene, 34

European Commission, 28, 51, 65, 68, 69, 71,
    79, 84, 87, 91
European Community, 25, 39, 40, 41, 51, 99,
    101
European Union, 52, 84
evaporation, 106, 109, 110, 114, 115, 118
evolution, 47, 61, 100, 121
exporter, 41
exports, 40
exposure, 18, 20, 82
extinction, 67
extra virgin oliv, vii, 9, 47, 52, 53, 59, 65, 66,
    70, 71, 74, 76, 77, 78, 79, 80, 82, 83, 91,
    95, 96, 116
extraction, 4, 9, 47, 48, 49, 53, 55, 56, 57, 58,
    59, 60, 62, 71, 75, 76, 80, 81, 82, 86, 89,
    90, 98, 99, 110, 111, 119
extraction process, 48, 49, 55, 59, 60, 75, 98

## F

family, 6, 11, 15, 25, 32, 48, 61
family members, 48
farmers, vii, 28
farming, 8
fat, 90, 99, 110
fatty acid, 6, 67, 68, 70, 72, 77, 81, 84, 85, 86,
    87, 88, 89, 90, 106
federal register, 68
feeding, 11, 120
fermentation, 25, 30, 34, 36, 37, 38, 41, 42,
    43, 57, 75, 99, 101, 102, 113, 115, 118, 121
fertility, 20, 107
fertilization, 21, 110
fertilizer, 5, 120
fiber, 57, 92, 100
filters, 108
filtration, 86, 108, 114
fixation, 18
flavor, 37, 66, 68, 70, 71, 73, 77, 85, 89
flocculation, 106, 108, 111, 115
flow, 34, 55, 73, 74, 109, 118

flue gas, 112
fluid, 58, 114
fluidized bed, 63, 111, 119, 121
fluorescence, 82
focusing, 99
food, 4, 5, 8, 9, 18, 27, 52, 65, 66, 79, 81, 91, 96, 99
food additives, 99
food products, 66
foodstuffs, 65
fractionation, 88, 119
fraud, 66, 79
fresh water, 113
frost, 19, 20
fructose, 102
fruits, 3, 7, 11, 25, 26, 27, 28, 29, 30, 32, 33, 34, 35, 36, 37, 38, 41, 42, 51, 52, 54, 73, 74, 90
fuel, 4, 101, 110
fungal, 105
fungi, 106

**G**

gas, 9, 34, 37, 38, 56, 78, 81, 89, 95, 101, 102, 109, 112
gas chromatograph, 81
gases, 102
gasification, 102, 110, 112
gasifier, 120
generation, 28, 34, 86, 101, 106, 112
germination, 103, 119
gibberellins, 13
gift, 3
glass, 38, 78
glucose, 36, 43, 102
glucoside, 32, 36
glycerol, 88
glycol, 104
glycoside, 25, 35, 37, 41
god, 4
gold, 4

government, iv
grades, 68, 83
grain, 47
grapes, 47
gravitational force, 58, 61
groundwater, 18
groups, 13, 27, 54, 83, 95, 96, 105, 107, 109, 110, 113, 115
growth, vii, 11, 13, 14, 15, 16, 18, 19, 20, 21, 36, 38, 100, 114, 120

**H**

handling, 54, 56, 74, 79
harvest, 15, 16, 52, 54, 59, 73, 81, 84
harvesting, vii, 9, 19, 30, 34, 51, 54, 73, 79, 97
hazards, 79
healing, 4
health, vii, 25, 26, 66, 79, 95, 96, 101, 121
heat, 30, 33, 34, 38, 55, 80, 101, 109, 110, 112, 113
heating, 30, 38, 74, 89, 98, 100
heavy metal, 107
Hebrew, 4
hemp fiber, 57
herbs, 28
high temperature, 85, 87, 88, 109
high-fat, 100
hormones, 13
horses, 3, 50
hot water, 48
human, 5, 65, 71, 95, 101, 103, 121
humanity, 4
humidity, 35, 70
humus, 100
hybrid, 12
hydration, 5
hydro, 85, 87, 112, 119
hydrocarbon, 85, 87, 90
hydrogen, 102, 106, 112, 115
hydrogen peroxide, 106, 115
hydrogenation, 83, 90, 91

hydrolysis, 18, 35, 36, 101
hydrophilic, 119
hydroxide, 34, 35, 72, 86, 89, 113

## I

identification, 9
immersion, 35
immigrants, 8
imports, 40
impurities, 72, 85
incidence, 37
incineration, 101, 106, 109, 110
income, 52, 84
indices, 76, 77
indigenous, 27
industrial, vii, 9, 35, 47, 49, 61, 88, 99, 102, 108, 110, 111, 118
industry, 42, 52, 118
inert, 38, 56, 78, 89
infection, 20
inflammatory, 96
inhibitory effect, 120
initiation, 14
injury, iv
inoculation, 36
inorganic, 16, 18, 102, 115
insects, 20
internal combustion, 112
investment, 101
ionization, 81, 117
ions, 18
iron, 49
irradiation, 107
irrigation, 21
island, 7, 8
isomerization, 90
isomers, 70, 86, 87, 88, 90

## J

Jews, 4

## K

kernel, 15, 52
kinetics, 17, 79, 117

## L

labor, 54, 59, 113
laccase, 121
lactating, 99
lactic acid, 36, 37
lactic acid bacteria, 36
Lactobacillus, 36, 42, 105
land, 8, 95, 100
landfill, 96
language, viii
leaching, 18
learning, 22, 82
legislation, 70, 73, 78
lignin, 14, 102
linear, 52
linoleic acid, 73
linolenic acid, 87, 88
lipase, 67, 71, 91
lipid, 63, 71, 81, 91, 92, 121
lipoprotein, 120
lipoxygenase, 71
liquid chromatography, 117
liquid phase, 60, 61
location, 110
logging, 17
long period, 28
losses, 18, 87

## M

machinery, 59
machines, 75
magnetic, iv
maintenance, 79
maize, 66

management, vii, 10, 21, 42, 62, 81, 84, 96, 99, 103, 116, 119

manufacturing, 5, 33, 41, 47, 48, 52, 57, 59, 61, 73, 79, 87, 96, 97, 102, 110, 113, 117

margarine, 83, 90, 91

maritime, 3, 8

market, 62

mass spectrometry, 81, 82, 117

mathematical methods, 7, 79

maturation, 15

maturation process, 15

media, 82

medicine, 5, 9

mediterranean, 4, 6, 7, 8, 9, 10, 27, 28, 41, 43, 49, 54, 65, 79, 85, 97, 119

melting, 85, 88

membranes, 109

metabisulfite, 35

metabolic, 110

metabolism, 19, 119

metabolites, 115

metals, 72, 85, 86, 107

methane, 101, 106

microbial, 36, 108, 122

microorganisms, 18, 36, 38, 71, 100, 104, 110

Middle East, 27

milk, 99

mineralization, 18

mitochondrial, 9

mixing, 56, 66, 75, 80

models, 78, 79

moisture, 58, 59, 74, 75, 98, 112, 121

moisture content, 74, 98

mold, 114

molecular mass, 117

molecules, 16, 17, 18, 88, 111

monks, 8

monomeric, 103

monounsaturated fatty acids, 90

multivariate, 82

mycelium, 121

mycenaean, 48, 49

# N

natural, 25, 27, 30, 31, 34, 37, 38, 41, 58, 59, 76, 85, 88, 96, 99, 100, 102, 105, 107, 113

near East, 7

negative influences, 73

network, 121

neural network, 81, 121

neutralization, 83, 85, 86, 89, 91

Newtonian, 58

nitrate, 18, 36

nitrogen, 13, 14, 16, 18, 19, 21, 56, 78, 100, 107

nitrogen fixation, 18

non-Newtonian fluid, 58

normal, 29, 58

North Africa, 8

nutrients, 11, 13, 15, 19, 21, 100

nutrition, 14, 66, 91, 96

# O

observations, 119

occupied territories, 49

odyssey, 4

oil mill waste waters, 107, 120

oil production, 8, 15, 21, 48, 49, 50, 51, 52, 62, 80, 99

oil refining, 91

oil samples, 72

olympic games, 4

operator, 76

optimal performance, 13

optimization, 101

organic, 15, 16, 18, 100, 101, 102, 105, 106, 107, 109, 113, 114, 115

organic compounds, 16, 106, 114, 115

organic matter, 100, 101, 105, 109

organoleptic, 31, 34, 52, 55, 65, 66, 70, 71, 73, 74, 75, 78, 84, 87, 91, 96, 115

osmosis, 114

osmotic, 120

OTC, 32
oxidation, 30, 31, 32, 41, 56, 59, 70, 71, 72, 74, 75, 76, 78, 85, 87, 106, 107, 109, 114, 115, 117, 119, 120, 121
oxidation products, 85, 87
oxidative, 67, 70, 71, 74, 76, 78, 79, 80, 86, 106
oxidative reaction, 67, 79
oxygen, 16, 17, 18, 59, 72, 78, 97, 98, 103, 105, 106, 109, 113, 114, 115
ozonation, 107, 111, 117
ozone, 114, 115

**P**

pacific, 28
pain, 5
palm oil, 90
parameter, 70, 71
parents, ix
particles, 86, 89, 108, 112
pasteurization, 38, 100
patents, 96
pathogenic, 107
PCA, 81
pectins, 15, 16
per capita, 39
perception, 72
percolation, 53, 57, 60, 61, 75, 76, 77
permeability, 35
permit, 68, 88
peroxide, 67, 70, 71, 73, 78, 84, 85, 106, 115, 117
pesticides, 74, 85, 87, 102
pests, 19
PET, 78
petroleum, 112
pH values, 102
pharmaceutical, 8, 9
phenol, 71, 76, 77, 107, 118
phenolic, 14, 35, 67, 71, 76, 99, 102, 104, 105, 106, 107, 114, 115, 116, 117, 119

phenolic compounds, 67, 76, 99, 102, 104, 105, 106, 107, 115, 116, 117, 119
phospholipids, 85, 86, 119
phosphor, 110
phosphorus, 16, 100, 107
photocatalysis, 107
photochemical, 18
photosynthesis, 16, 17, 18
physicochemical, 36, 38, 65, 66, 74, 78, 91, 113, 115, 116
physics, 18
physiological, 18, 21
pigments, 15, 72, 85, 86
plants, 4, 11, 13, 18, 20, 26, 49, 100, 103, 120
plastic, 37, 38, 54, 58, 74, 78, 90
pleurotus ostreatus, 121
plums, 15
pollen, 7, 15
pollutants, 97, 107
pollution, 97, 114
polycyclic aromatic hydrocarbon, 112
polymerization, 37, 38
polymorphisms, 9
polyphenolic compounds, 115
polyphenols, 42, 58, 61, 72, 75, 76, 89, 98, 102, 103, 105, 106, 113, 114
polyunsaturated fatty acids, 90
pomace, 52, 58, 59, 63, 66, 68, 69, 71, 83, 84, 87, 88, 90, 95, 96, 97, 98, 110, 111, 112, 116, 121
poor, 85, 99
population, 4, 7, 15, 39, 49
pore, 115
porosity, 96, 99
porous, 108
potassium, 16, 72, 107, 110
pouches, 38
power, 4, 5, 8, 39, 47, 57, 108, 112
precipitation, 108, 111
preference, 33
present value, 36
preservatives, 38

pressure, 30, 48, 53, 57, 60, 61, 75, 76, 97, 103
prevention, 56
prices, 52, 66, 84
principal component analysis, 80
processing stages, 28
production technology, 49
productivity, vii, 9, 13, 20, 39, 49, 52, 58, 59, 62
property, iv, 116
propionic acid, 36
protection, 78
proteins, 13, 18, 41, 85, 99, 102
pruning, vii, 7, 8, 11, 13, 16, 18, 19, 20, 21, 54
purchasing power, 39
purification, 4, 92, 114, 117, 119
pyrolysis, 101, 110

relevance, 70
reliability, 110
remediation, 105, 110
reparation, 31, 41, 92
residues, 18, 87, 95, 99, 101, 116
resistance, 38, 74, 75, 76, 79
respiration, 16, 17, 18, 34, 37
returns, 18
ripeness, 15, 30, 31, 32, 54, 55, 73
risks, 20, 101
rolling, 47, 50
roman empire, 49, 57
room temperature, 56, 74
root cap, 13
ruminant, 120
runoff, 18

## Q

quantization, 22, 82

## R

race, 85
radiation, 115
radius, 12
random, 32
range, 16, 17, 84, 96, 97, 102
raw material, viii, 8, 58, 109
reactant, 37, 102
reaction rate, 17
reagent, 17, 115, 119
recovery, 121
recycling, 58, 114, 115, 121
refining, 69, 84, 85, 86, 88, 89, 91
regeneration, 114, 121
regional, 51
regression, 82
regular, 19
regulations, 68, 84, 85
relationship, 7, 11, 15, 56

## S

sacred, 4
saline, 9, 28
salinity, 107
salt, 28, 30, 33, 37, 113, 114
sand, 108
sanitation, 73
saturated fatty acids, 77, 88
scaling, 28, 89
scientific knowledge, 74
seawater, 28
sedimentation, 38, 53, 61, 77, 88, 109
seed, 5, 8, 9, 47, 48, 66, 84, 91, 103, 119
seedling development, 103
seedlings, 11
self, 22, 82
semiarid, 13
separation, 52, 57, 95, 108, 109, 114
services, iv
sewage, 103
shape, 6, 31
shoot, 54
short period, 38, 107
shrubs, 6

sign, 4
sites, 7
skeleton, 19
skin, 5, 15, 31, 35, 52, 57
sludge, 87, 101, 105, 106, 109, 117
sodium, 34, 35, 36, 37, 38, 43, 86, 89, 113
sodium hydroxide, 34, 35, 86, 89, 113
soil, 13, 14, 16, 18, 20, 21, 54, 100, 101, 105, 107, 119, 120, 121
solar, 16
solar energy, 16
solid phase, 57, 88
solid waste, 63, 97, 100, 101, 102, 118, 121
solvents, 8, 72, 86, 90
sorghum, 103, 104
sorting, 34
South Africa, 9, 27
South America, 9
soybean, 66
species, 6, 11, 15, 19, 25, 39, 105, 120
spectrophotometric, 73
spectrophotometry, 81
spectroscopy, 82, 86
spectrum, vii
speed, 58
spheres, vii, 61
stability, 38, 70, 75, 76, 77, 78, 80, 85
stabilize, 38, 110
stages, 14, 28, 33, 36, 37, 41, 49, 56, 73, 79, 83, 84, 85, 113, 114
stainless steel, 56, 60, 78
stamens, 15
standards, 72, 82
starch, 13, 16
steel, 56, 60, 78
sterile, 12
sterilization, 30, 38
sterols, 86, 87
storage, 13, 18, 34, 36, 37, 41, 42, 43, 54, 70, 71, 72, 74, 78, 79, 80, 81, 82, 97, 101, 102
strain, 58
strategies, 11
streams, 113

strength, 4
stress, 14, 20
students, vii
substances, 14, 72, 86, 89, 102, 107, 108, 112, 115, 117
substitutes, 66
substitution, 66
sucrose, 102
sugar, 16, 18, 26, 36, 42, 92, 110
sugar beet, 92
sulfur, 85, 101, 110
sulphate, 106
summer, 13, 14, 20
sunflower, 66
sunlight, 16, 20
supercritical, 83, 89, 91
supercritical carbon dioxide, 89, 91
supply, 8, 17, 100
surface area, 13, 59, 99
syndrome, 66
synthesis, 13, 95, 102, 112
synthetic fiber, 57
systems, 21, 35, 50, 60, 61, 62

# T

talc, 75
tandem mass spectrometry, 117
tanks, 61, 77, 88
tannins, 102
tar, 102
taste, 25, 26, 27, 28, 35, 37, 41, 55, 65, 66, 73, 76, 87
technology, 9, 21, 42, 47, 49, 57, 61, 62, 80, 91, 101, 109, 110, 115
temperature, 16, 17, 21, 34, 35, 37, 38, 54, 55, 56, 59, 72, 74, 75, 77, 79, 87, 100, 114
thermal resistance, 38
thiamine, 41
ticks, 60

time, vii, 3, 4, 7, 12, 26, 27, 30, 35, 36, 37, 38, 56, 58, 61, 67, 72, 74, 78, 79, 81, 89, 102, 107, 112
timing, 20
tin, 38, 78
tocopherols, 77, 87
tolerance, 9
toxic, 66, 97, 100, 102, 106, 111, 115
toxicity, 102, 106, 118, 120
trace elements, 89
trade, 3, 4, 8, 9, 25, 27, 28, 30, 31, 32, 39, 41, 48, 52, 61, 68, 84
Trade Act, 25
trading, 27, 68
training, 21, 62
trans, 86, 87, 88, 90
transformation, 16, 100
transition, 14
transport, 4, 54, 74, 79
transportation, 34, 35, 101
treatment methods, 105
trees, 3, 4, 6, 7, 8, 9, 11, 12, 13, 16, 17, 19, 20, 21, 27, 51, 54, 99
triacylglycerols, 80, 85, 88, 92
trial and error, 28
triglyceride, 67
tubular, 102

values, 19, 36, 38, 39, 57, 66, 70, 71, 72, 75, 84, 85, 86, 87, 99, 102
variability, 37
variables, 66, 75
variation, 10, 43
vector, 22, 82
vegetable oil, 22, 81, 82, 92
vegetables, 42
vegetation, 110, 119
ventilation, 54, 74
vitamins, 102
volatilization, 18

# U

ultraviolet, 67
uniform, 30
United Nations, 28, 43, 68
USDA, 67
UV irradiation, 107
UV radiation, 115

# V

vacuum, 38, 87, 89, 110, 115

# W

war, 4
waste treatment, 101, 114
waste water, 58, 59, 76, 95, 96, 97, 100, 102, 103, 105, 106, 107, 108, 109, 113, 114, 115, 116, 117, 118, 119, 120, 121
wastes, vii, 95, 96, 99, 100, 101, 102, 112, 113, 116, 118
wastewater treatment, 120
waxes, 14, 85, 88
wealth, 4
web, 42
wildlife, 11
wine, 5, 30, 47, 48
winter, 20, 54
wisdom, 3
wood, 9, 12, 20, 28, 38

# Y

yeast, 114
yield, vii, 57, 59, 102